Introduction to Chaos, Fractals and Dynamical Systems

Problem Solving in Mathematics and Beyond

Print ISSN: 2591-7234
Online ISSN: 2591-7242

Series Editor: Dr. Alfred S. Posamentier
Distinguished Lecturer
New York City College of Technology - City University of New York

There are countless applications that would be considered problem solving in mathematics and beyond. One could even argue that most of mathematics in one way or another involves solving problems. However, this series is intended to be of interest to the general audience with the sole purpose of demonstrating the power and beauty of mathematics through clever problem-solving experiences.

Each of the books will be aimed at the general audience, which implies that the writing level will be such that it will not engulfed in technical language — rather the language will be simple everyday language so that the focus can remain on the content and not be distracted by unnecessarily sophiscated language. Again, the primary purpose of this series is to approach the topic of mathematics problem-solving in a most appealing and attractive way in order to win more of the general public to appreciate his most important subject rather than to fear it. At the same time we expect that professionals in the scientific community will also find these books attractive, as they will provide many entertaining surprises for the unsuspecting reader.

Published

Vol. 29   *Introduction to Chaos, Fractals and Dynamical Systems*
            by Phil Laplante and Chris Laplante

Vol. 28   *Seduced by Mathematics: The Enduring Fascination of Mathematics*
            by James D Stein

Vol. 27   *Mathematics: Its Historical Aspects, Wonders and Beyond*
            by Alfred S Posamentier and Arthur D Kramer

Vol. 26   *Creative Secondary School Mathematics: 125 Enrichment Units
            for Grades 7 to 12*
            by Alfred S Posamentier

For the complete list of volumes in this series, please visit www.worldscientific.com/series/psmb

Problem Solving in
Mathematics and Beyond | Volume **29**

Introduction to Chaos, Fractals and Dynamical Systems

Phil Laplante

Penn State University, USAA

Chris Laplante

Agilent Technologies, USA

 World Scientific

NEW JERSEY · LONDON · SINGAPORE · BEIJING · SHANGHAI · HONG KONG · TAIPEI · CHENNAI · TOKYO

Published by

World Scientific Publishing Co. Pte. Ltd.
5 Toh Tuck Link, Singapore 596224
USA office: 27 Warren Street, Suite 401-402, Hackensack, NJ 07601
UK office: 57 Shelton Street, Covent Garden, London WC2H 9HE

Library of Congress Cataloging-in-Publication Data
Names: Laplante, Phillip A., author. | Laplante, Chris, author.
Title: Introduction to chaos, fractals and dynamical systems / Phil Laplante
    (Penn State University, USA), Chris Laplante (Agilent Technologies, USA).
Description: New Jersey : World Scientific, [2024] | Series: Problem solving in mathematics and
    beyond, 2591-7234 ; vol. 29 | Includes bibliographical references and index.
Identifiers: LCCN 2023015059 | ISBN 9789811273247 (hardcover) |
    ISBN 9789811273902 (paperback) | ISBN 9789811273254 (ebook for institutions) |
    ISBN 9789811273261 (ebook for individuals)
Subjects: LCSH: Fractals. | Dynamics. | Chaotic behavior in systems.
Classification: LCC QA614.86 .L368 2024 | DDC 006.601/514742--dc23/eng20230711
LC record available at https://lccn.loc.gov/2023015059

British Library Cataloguing-in-Publication Data
A catalogue record for this book is available from the British Library.

For any available supplementary material, please visit
https://www.worldscientific.com/worldscibooks/10.1142/13331#t=suppl

Desk Editors: Sanjay Varadharajan/Ana Ovey

Typeset by Stallion Press
Email: enquiries@stallionpress.com

Printed in Singapore

For Kaitlin

About the Authors

Phillip A. Laplante started playing with computers in the 1970s. By his senior year in college, he was writing software for the Space Shuttle and other avionics applications. Later, he worked on computer-aided design software for microwave and electronics systems and then on advanced testing tools. While working on these projects, he completed his Master's degree in electrical engineering and his PhD in computer science. He is currently Professor of Software and Systems at Penn State where his teaching and research interests have included image processing, real-time systems, software testing, artificial intelligence, critical infrastructure and the Internet of Things. You can find out more about his other books and publications at https://phil.laplante.io.

Christopher P. Laplante was introduced to the BASIC programming language at a very young age by his father, Phil. He quickly became fascinated by programming and computers in general. He is currently a software engineer at Agilent Technologies where he works on embedded Linux systems for gas chromatographs and automated liquid samplers. You can learn more about his side projects at https://blog.laplante.io.

Acknowledgments

We would like to thank the following individuals:

- our acquisition editor, Rochelle Kronzek, for commissioning this project and support and encouragement along the way;
- Dr. Alfred S. Posamentier for including our book in his wonderful *Problem Solving in Mathematics and Beyond* book series.

We'd also like to acknowledge the following friends for reviewing portions of the text at various stages of draft:

- Dr. Joanna DeFranco, Penn State
- Dr. Mohamad Kassab, Penn State
- Len Taddei, Agilent Technologies
- Steve Kirk, Agilent Technologies

Of course, any remaining errors or inconsistencies are our own.

We hope you enjoy this exploration of chaos, fractals and dynamical systems. If you think you have found an error in the book and would like to report it, please contact us via email at:

Phillip A. Laplante, *plaplante@psu.edu*

Christopher P. Laplante, *chris@laplante.io*

Contents

About the Authors   vii

Acknowledgments   ix

List of Figures   xv

List of Tables   xix

Introduction   xxi

1. What is Chaos? What are Fractals?   1

   1.1   Stable/Unstable Systems   1

   1.2   What is Chaos?   2

   1.3   What are Fractals?   3

   1.4   Other Fractal-like Things   5

   1.5   How are Fractals Created?   5

        1.5.1 Dynamical Systems   5

        1.5.2 Algorithms   6

        1.5.3 Attracting and Escaping Points   7

        1.5.4 Bifurcation Diagrams   9

   1.6   The Sierpinski Triangle   13

   1.7   Iterated Function System Transformations   18

   1.8   Recursive Generation of Fractals   25

        1.8.1 The Cantor Set   28

   1.9   Fractal Dimension   31

   1.10 How are Fractals and Chaos Related?   33

1.11 Brief History of Chaos, Fractals and
       Dynamical Systems 34
1.12 Exercises and Things to Do 35

2. Foundations of Chaos and Fractal Theory   39
    2.1 Complex Numbers and Functions 39
         2.1.1 Plotting Complex Numbers 40
         2.1.2 Arithmetic with Complex Numbers 41
    2.2 Functions of Complex Variables 44
    2.3 Generating Fractals by Finding Attractors
         of Complex Functions 46
         2.3.1 Julia Sets 46
         2.3.2 The Histogram Coloring Algorithm 52
         2.3.3 More Julia Sets 57
         2.3.4 The Mandelbrot Set 61
         2.3.5 A Note on the Images 64
         2.3.6 The Inverse Iteration and Boundary
                Scanning Methods 65
    2.4 Three-Dimensional Fractals 67
    2.5 Wavelets 68
    2.6 Exercises and Things to Do 71

3. Chaos and Fractals in Nature   73
    3.1 Population Dynamics 73
    3.2 Animal Images 78
    3.3 Genetics 79
    3.4 Weather 79
    3.5 Scenes from Nature 81
         3.5.1 Trees, Leaves and Flowers 81
         3.5.2 Clouds 90
         3.5.3 Rocks and Boulders 92
         3.5.4 Snowflakes 95
         3.5.5 Galaxies 96
         3.5.6 Coastlines 98

3.6 Fractals in the Human Body 99
   3.6.1 Bronchial Growth 100
   3.6.2 Neuron Growth 100
   3.6.3 Physiological Processes 102
   3.6.4 Chaos of the Mind? 103
3.7 Exercises and Things to Do 103

4. Chaos and Fractals in Human-Made Phenomena   107

4.1 Turbulent Flow 107
4.2 Structures 108
4.3 Computer Scene Analysis 111
4.4 Image Compression 111
   4.4.1 Problems with Fractal Compression 113
4.5 Economic Systems 115
4.6 Cellular Automata 118
   4.6.1 One-Dimensional Cellular Automata 119
   4.6.2 Two-Dimensional Cellular Automata 124
4.7 Exercises and Things to Do 127

5. Dynamical Systems and Systems Theory   129

5.1 Basic System Theory 130
   5.1.1 Modeling Things as Systems 131
   5.1.2 Linear Growth 133
   5.1.3 Response Time 136
   5.1.4 Control Systems 136
   5.1.5 Deterministic Systems 139
   5.1.6 Time-Variant and Time-Invariant Systems 140
   5.1.7 Linear Systems 140
   5.1.8 Nonlinear Systems 143
   5.1.9 Summary of System Types 144
5.2 Systems Thinking 144
   5.2.1 Emergent Behavior 145
   5.2.2 Second-Order Effects 147

5.3   Uncertainty 148
       5.3.1   State Uncertainty 148
       5.3.2   Heisenberg Uncertainty 149
       5.3.3   Uncertainty and Chaos 150
       5.3.4   Dealing with Uncertainty 150
5.4   Swarm Behavior 155
       5.4.1   Simulating a Swarm 157
       5.4.2   Applications of Simulated Swarms 158
       5.4.3   Swarm Computing 159
5.5   Brief History of Dynamical Systems and
       Systems Theory 160
       5.5.1   Dynamical Systems 160
       5.5.2   Systems Engineering 161
       5.5.3   Origins of Systems Thinking 161
5.6   Exercises and Things to Do 161

Glossary   165

Bibliography   179

Index   181

List of Figures

1.1   Two sections of the infinite roller coaster. (a) Stable equilibrium; (b) Unstable equilibrium. 2

1.2   Increasingly nearer views of a section of coastline. 4

1.3   The Mandelbrot set. 4

1.4   Bifurcation diagram for $f(x) = x^2 + c$ with $x = 0$ and c swept from -2.0 to 1.0. 10

1.5   Rule for creation of a Sierpinski triangle. 13

1.6   A Sierpinski triangle. 14

1.7   Sierpinski triangle-mapping procedure. 15

1.8   Vertices for Sierpinski triangle-mapping procedure. 16

1.9   Starter triangle. 18

1.10 The original triangle (dashed) scaled to 50% scale. 19

1.11 Translations of scaled triangle. 19

1.12 First iteration of Sierpinski triangle. 20

1.13 Sierpinski carpet. 24

1.14 Affine transformations for constructing first iteration of Sierpinski carpet. Original square (dashed) is scaled to 33% scale. Then it is cloned and translated in eight different directions. 25

1.15 Construction of the Cantor set. 28

2.1   The complex plane and some points on it. 41

2.2   The plotting window for Julia and Mandelbrot programs. . . . 49

2.3   Julia set for $f(z) = cos(z)$. 50

2.4   Douady's rabbit. 51

2.5   Inferno color scheme.   52

2.6   Douady's rabbit created using 10 iterations.   53

2.7   Inferno color scheme utilization (dashed box) for Douady's rabbit with 10 iterations. Note how the darker colors on the left are not used because the escape iteration count starts at 3.   53

2.8   Douady's rabbit created using 50 iterations.   54

2.9   Inferno color scheme utilization (dashed box) for Douady's rabbit with 50 iterations.   54

2.10 Douady's rabbit created using 200 iterations.   55

2.11 Douady's rabbit created using 800 iterations.   56

2.12 Douady's rabbit created using 800 iterations with histogram coloring.   57

2.13 A Siegel disk.   58

2.14 A dragon.   59

2.15 The Julia set of sin(z).   60

2.16 A filled Mandelbrot set.   63

2.17 The Mandelbrot set.   64

2.18 The Mandelbrot set using the histogram coloring algorithm.   65

2.19 Basin of attraction used in BSM.   66

2.20 Mexican hat function mother wavelet.   69

2.21 Some daughter wavelets of the Mexican hat function.   70

2.22 Simulated electrocardiogram image produced by Mexican hat wavelets.   70

3.1   Population dynamics of caribou–wolf system with $K = 0.000006$.   76

3.2   Population dynamics of caribou–wolf system with $K = 0.000010$.   76

3.3   Population dynamics of caribou–wolf system with $K = 0.000014$.   77

3.4   Seals.   78

3.5   An amoeba-like image generated from the filled Julia set of $f(z) = Z^2 + 0.3 - 4i$.   80

3.6   A fern leaf.   81

3.7   A fern leaf generated via IFS. 82

3.8   IFS representation of a tree. 84

3.9   A forest of randomly generated fractal trees. 85

3.10 A redwood forest. 88

3.11 Green seaweed. 89

3.12 Four-petaled flower from the Julia set of
      $f(z) = z^2 + 0.384.$ 90

3.13 Another four-petaled flower from the Julia set of
      $f(z) = z^2 + 0.2541.$ 91

3.14 A fractal storm cloud. 91

3.15 "Three-dimensional" fractal clouds. 93

3.16 Fractal rocks using same IFS codes as the clouds. 94

3.17 Snowflakes generated by mandel_julia.rs. 95

3.18 A snow fall generated by fall.rs. 97

3.19 Cross fractal used to generate snowflakes. 98

3.20 A randomly generated view of space. 99

3.21 The bronchial tree structure in our lungs resembles
      a fractal. 100

3.22 Dendrite structure generated by $f(z) = z^2 + i.$ 101

3.23 An EKG output simulated using a Julia set. 102

4.1   Castle. 109

4.2   A fractal maze. 110

4.3   "Winter reflections" by waferboard is marked with
      CC BY 2.0. To view the terms, visit https://creativecommons.
      org/licenses/by/2.0/?ref=openverse. Originally in color. . . . 113

4.4   Computer-generated equivalent of swampy pond shown in
      "winter reflections." 114

4.5   Self-similarity of cotton prices over centuries, years,
      and months. 116

4.6   Bifurcation diagram for model economic system. 117

4.7   Sierpinski triangle output of program 1d_life.rs. 121

4.8   Sample output from 1d_life.rs using Rule2. 123

4.9   Game of Life. 125

4.10 Game of Life after many iterations consisting of just stable and
       bi-stable formations of cells. 126
5.1   A general system with inputs and outputs. 130
5.2   A system with one input and one output. 131
5.3   Control system block diagram. 137
5.4   A real system interacts with the environment in obvious
       and subtle ways. 145
5.5   Honeybee swarm near Wilmington, Delaware, courtesy of
       Small Wonder Honey, https://www.facebook.com/smallwonde
       rhoney. 156
5.6   Screen capture of an artificial swarm using Wolfram's Boids
       Simulator at https://demonstrations.wolfram.com/BoidsSimul
       atedFlockingBehavior. ©Wolfram Research. 158

List of Tables

1.1   IFS transformation rule for Sierpinski triangle.   20
1.2   IFS transformation rule for Sierpinski carpet.   23

3.1   IFS transformation rule for seals.   79
3.2   IFS transformation rule for fern.   83
3.3   IFS transformation rule for tree.   83
3.4   IFS transformation rule for seaweed.   90
3.5   IFS codes for three-dimensional fractal clouds.   93
3.6   IFS codes for cross fractal.   98

4.1   IFS codes for castle.   109
4.2   IFS codes for maze.   111
4.3   IFS codes for one clump in a swamp.   115

5.1   Principal and interest for two years in a savings account. . . .   134
5.2   Simplified interest growth model.   134

Introduction

1. Background and Objectives

Thank you for choosing this book. Our goal is to provide a fun and enriching introduction to chaos theory, fractals and dynamical systems. An emphasis is made on natural and human-made phenomenon that can be modeled as fractals and on the applications of fractals to computer-generated graphics and image compression.

To keep the mathematics relatively simple, we rely on intuitive descriptions, computer-generated graphics and photographs of natural scenes to make our points. But there are sufficient mathematical definitions, representations, discussions and exercises so that this book can be used as primary or secondary source in home schooling environments. We also present a brief history of the evolution of chaos theory, fractals and dynamical systems. Again, just enough information is given to intrigue the reader and get them started on other scientific journeys.

To indulge curious minds (both ours and the readers'), we've included numerous factoids, biographies and short discussions in footnotes. We don't intend these as distractions, but rather to encourage further inquiry. In some cases, these footnote items reappear at the end of the chapter as exercises and things to do. We also tried to keep the citations to a manageable level. We didn't intend this to be referenced as a scholarly paper. Rather, we give just enough citations for those interested in conducting further research.

2. Mathematical Background

The book is accessible to those with a basic knowledge of algebra and geometry. In particular, the reader should be familiar with simple functions of real numbers and it would also be helpful if the reader were familiar with a little trigonometry, such as sines and cosines. However, much of the needed mathematical background is developed along the way. In a few places, we expand somewhat into higher mathematics, including the use of a little calculus. In these cases, however, the mathematics can be ignored or used as the staring point for further exploration.

The mathematics is introduced in a very informal (that is, not rigorous) way. We realize that in some cases (e.g. the definition of a real number, linear system, fractal) we have taken liberties. But to do otherwise would require a higher level of sophistication on the part of the reader and might deter them from further study. We'd rather give the more informal definitions, letting the reader (or teacher or professor) introduce further rigor to the level desired.

3. Rust and OpenGL

While the book can be enjoyed without a computer, we hope that the reader will experiment with the code samples provided — these are available at https://code.laplante.io for download. We chose to implement the various fractals and other programs in the Rust programming language. Our reasons for doing so are detailed in the following section. Please keep in mind that this book is not intended to be an introduction or tutorial on Rust, nor programming in general. That being said, we have tried our best to explain the code that appears in this book.

3.1 *The Rust Program Language*

Rust is a programming language with syntax similar to C/C++ but incorporating features more commonly found in the ML-family of languages, such as algebraic data types and pattern matching. For professional software developers, Rust's most important feature is compile-time safety. The Rust compile-time checker looks for null pointers, dangling references, data races and other kinds of undefined and unsafe behavior using a variety of techniques including a "borrow checker," which analyzes the lifetimes of objects and references looking for anomalies. Since the borrow checker runs at compile time, the generated assembly code performs similar to the code written in C and C++ (Langlands *et al.*, 2021).

We primarily chose Rust for this book because it provides an excellent developer experience. It is simple to install with rustup (https://rustup.rs/) and runs on Windows, Linux and macOS. The built-in package manager and build system, cargo, makes it easy to build software. It's also very reliable, as mentioned previously. Rust has great documentation and a very user-friendly compiler — the error messages are not cryptic unlike many other programming language implementations. For more information on Rust, check out https://www.rust-lang.org. There is an excellent online free Rust book here: https://doc.rust-lang.org/book.

A 1993 predecessor to this book contained code written in Pascal, which was a popular teaching language at the time. Code for all of the programs in the book was also made available in C. The legacy versions of the code in C can be downloaded from https://github.com/introtochaosbook/c-fractal-code. These will need to be modified to run in your particular programming computing environment if you wish to run the code.

3.2 *OpenGL and Hardware-Accelerated Graphics*

For the most part, the Rust code is merely a wrapper around one or more OpenGL programs. The OpenGL programs are responsible for determining what will be drawn to the screen. OpenGL ("Open Graphics Library") is a standardized API for manipulating graphics hardware. OpenGL programs are called *shaders* and are written in a C-like language called GLSL.[1]

---

[1]OpenGL Shading Language.

Shaders are compiled into instructions that run directly on GPUs and other graphics hardware (Kessenich *et al.*, 2017).

It may seem like overkill to use both Rust and OpenGL, but we have good reasons for doing so. While we certainly could have written programs entirely in Rust that draw fractals on the screen, the resulting code would be extremely slow and inefficient.[2] This is because such programs would need to iterate over every pixel[3] on the screen and perform a calculation — a daunting task for today's densest 4K or 8K monitors. By using OpenGL, we offload processing from the CPU (which is not particularly optimized for any one task) to the GPU (which was purpose-built for drawing to the screen).

Another benefit of using OpenGL is that since the "meat" of the code is written in GLSL, you can easily port the fractals to other programming languages by simply rewriting the Rust wrapper code. OpenGL bindings are available for most popular programming languages and platforms. Tutorials and project templates are also readily searchable online.

For readers that are not familiar with OpenGL, we recommend the free website, https://learnopengl.com. The code samples are in C++, but the concepts are the same. The following sections cover the bare minimum OpenGL knowledge required to understand the example programs.

3.3 *OpenGL Coordinate System*

The OpenGL coordinate system is rather simple. Regardless of screen width and height, x- and y-coordinates range from -1.0 to 1.0. See the following figure for an illustration of the OpenGL coordinate system:

---

[2]For the sake of simplicity, this *is* the approach we took for some fractals, specifically the iterated function system (IFS) class of fractals.

[3]This pixel-by-pixel approach is how the fractals in the aforementioned 1993 predecessor to this book were implemented. For some historical context, OpenGL was first released in 1994, but the first commercially available GPU (the NVIDIA GeForce 256) wasn't widely available until 1999.

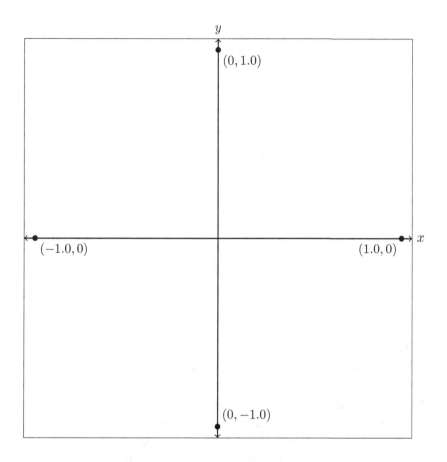

3.4 *Vertices, Shaders and the OpenGL Render Pipeline*

OpenGL specifies a handful of different shader types. We will be using three: vertex, geometry, and fragment shaders.[4] The shaders are composed together to form a render pipeline which controls how OpenGL renders a scene. The OpenGL render pipeline is depicted in the following figure:

---

[4]For the full list, see https://www.khronos.org/opengl/wiki/Shader.

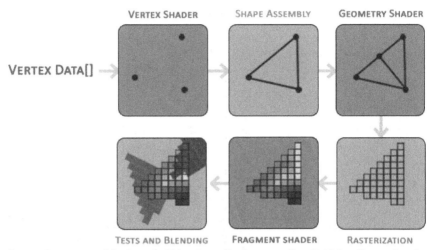

Source: Image copyright Joey de Vries, https://twitter.com/JoeyDeVriez, licensed under CC BY 4.0 (https://creativecommons.org/licenses/by/4.0/). From article https://learn opengl.com/Getting-started/Hello-Triangle. Retrieved from https://learnopengl.com/img/getting-started/pipeline.png.

The rendering process begins with us feeding a list of vertices (points in 3D space) to the GPU. For each vertex, the *vertex shader* is run. Vertex shaders manipulate the 3D coordinates of each vertex, one at a time. One common usage for vertex shaders is to implement camera transforms. Our programs won't actually need to perform any vertex processing, but OpenGL still requires a vertex shader. We will simply use a pass-through vertex shader that returns each input vertex unchanged:

```
#version 140
in vec2 position; // (x,y)
void main() {
    // output (x, y, z, depth) with z = 0.0, depth = 1.0
    gl_Position = vec4(position, 0.0, 1.0);
}
```

Vertices are then assembled into primitive shapes. Assembly requires that we tell OpenGL how to interpret the vertices. Most of our example programs use triangles, in which case OpenGL consumes the vertices three at a time and interprets them as the endpoints of a triangle. Nearly all the example programs use two triangles (six vertices in total) that when combined form a square that covers the whole screen. In code form, the vertices look like this:

```
let vertices: [Vertex; 6] = [
    // first triangle
    [1.0, -1.0].into(),
    [-1.0, 1.0].into(),
    [-1.0, -1.0].into(),
    // second triangle
    [1.0, 1.0].into(),
    [1.0, -1.0].into(),
    [-1.0, 1.0].into(),
];
```

Refer to the earlier figure, showing an illustration of the *OpenGL coordinate system*, to convince yourself that these two triangles cover the whole screen.

Assembled shapes are then sent to the *geometry shader* if present. Geometry shaders have the ability to emit new vertices. The only time we will need geometry shaders is for the first example program.

After the geometry shader, the final set of primitives is mapped into the set of pixels that will be displayed on the screen (i.e. rasterized). The *fragment shader* runs on each of these pixels. Fragment shaders are where most of the interesting work is done in our programs. They are responsible for determining the final color of each pixel.

Finally, OpenGL performs per-sample operations that can include certain tests and blending operations. These will be ignored for our purposes.[5]

4. Running the Programs

If you'd like to run any of the fractals for yourself, you will need to download (clone) the source code from our GitHub page: https://code.laplante.io. You will also need to install Rust and a few other packages depending on your operating system. We will keep updated instructions for setup on that page.

---

[5]If these interest you, see https://www.khronos.org/opengl/wiki/Per-Sample_Processing.

Throughout the book, you will see boxes that look like this:

```
Command line
cargo run --bin mandelbrot_simple
```

These boxes give the particular command that you should type in your terminal to run the fractal you see on the page.

5. Organization

This book is suitable for homeschooling as a focused course on the subject matter or as a classroom supplement for a variety of courses at the late junior high or early high-school level. For example, in addition to a standalone course on Chaos, Fractals and Dynamical Systems (or similar title), this book could be used with the following courses:

- Precalculus
- Geometry
- Computer programming (e.g. Rust, C, C++, Python, Java, Pascal)
- Computer graphics

The text can be used in conjunction with mathematics courses for undergraduates who are not science majors. The book can also be used for informal family study and discussion.

For each chapter, we've included exercises and things to do that range from simple computational tasks to more elaborate computer projects, related activities, biographical research and writing assignments. We encourage the use of Wolfram Alpha for confirming the answers to the computational exercises and for further experimentation. Wolfram Research's Web site, www.wolfram.com, provides many opportunities for playful experimentation.

6. About the Images

Since the book is printed in grayscale, we could not include color images — and you are really missing out on some of the beauty of the fractals and various fractal-like pictures. However, color files for all images in the book can be found at our website at https://fractals.laplante.io. For your convenience, we have added QR codes next to each image that will take you to the full-color versions.

They look like this: *Full-color image:*

Except where otherwise noted, the copyrights to all images, drawings, figures and code used in the book are held by Christopher Laplante and used with permission in this book. Use of any of these images, drawings, figures and code is prohibited unless explicit permission is obtained.

7. Our Wish

We hope you find the study of chaos, fractals and dynamical systems to be informative, entertaining and timely. After all, in the complex world in which we live, these phenomena are found everywhere, and we hope to enable readers to recognize these instances to enjoy and understand them.

Chapter 1

What is Chaos? What are Fractals?

"First there was Chaos, the vast immeasurable abyss,
Outrageous as a sea, dark, wasteful, wild."

— John Milton, *Paradise Lost*

In this chapter, we introduce the concepts of chaos, instability, stability, and other ideas that relate to fractals. We define a fractal, see some early examples, and look at the history of fractals, chaos, and start a brief foray into the mathematical field of dynamical systems, where they are studied.

1.1 Stable/Unstable Systems

Consider the section of infinite roller coaster shown in Fig. 1.1(a).

The car is located in the trough and is still. If we shove the car gently in either the front or the back, it is clear that friction, gravity, and rotational kinetics will act to return the car to the trough. In fact, within certain limits, it does not matter how far the car is pushed in either direction — the car consistently returns to the same place. This system is said to be in *stable equilibrium*.

Now, consider the section of the same roller coaster shown in Fig. 1.1(b). If you were to gently nudge the roller coaster in the front or the back, the car begins a wild ride and it is unclear where the car stops. This system is said to be at *unstable equilibrium*. In common language, we might refer to the position of the car on this roller coaster as a "tipping point." The concepts of stable and unstable equilibrium, as well as sensitivity to initial conditions

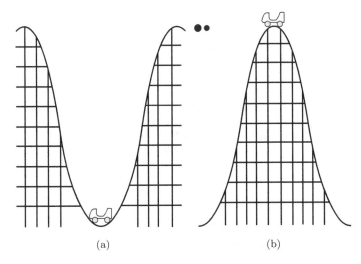

Fig. 1.1. Two sections of the infinite roller coaster. (a) Stable equilibrium; (b) Unstable equilibrium.

(in this case where the car starts), are crucial in the study of fractals, chaos and dynamical systems.

1.2   What is Chaos?

Chaos is derived from a Greek verb which means "to gape open," but in our society, chaos evokes visions of disorder. In a sense, chaotic systems are those that when they are in equilibrium, they are in unstable equilibrium — even the slightest change to the initial conditions of the system at time t leads the system to a very different outcome at some arbitrary later time. Such systems are said to have a *sensitive dependence* on initial conditions.

Some system models, such as that for the motion of planets within our solar system, contain many variables, yet still are accurate. With chaotic systems, however, even when there are hundreds of thousands of variables involved, no accurate prediction of their behavior can be made.

For example, the weather is known to be a chaotic system — despite the best efforts of beleaguered meteorologists to forecast, they very frequently err. This cause–effect pair is typical of a chaotic system and illustrates sensitive dependence on initial conditions.

You encounter chaotic systems in other aspects of your life. Traffic patterns tend to be chaotic — the errant maneuver of even one driver can create an accident or traffic jam that can affect thousands of others. The behavior of crowds can appear chaotic. Many people feel that the stock market is a chaotic system because the behavior of one investor, political situation, corporation, and so forth can affect billions of dollars in investments (Peters, 1991). Finally, those of you that enjoy science fiction are familiar with story lines where a time traveler goes back and alters a course of events, even slightly, with traumatic consequences.[1] Just as the ripples from a pebble tossed into a lake affect the farthest shore, our slightest actions can have far-reaching repercussions.[2]

1.3 What are Fractals?

There is a rigorous and precise mathematical definition of a *fractal*, but it is beyond the scope of this text. For our purposes, a fractal[3] is an image with an infinite amount of self-similarity. We will see many examples of fractals in this book, but fractals can be found just about anywhere you look, and we hope to raise your awareness of these possibilities.

But what is self-similarity? In natural and human-made phenomenon, *self-similarity* means that the structure of the whole is often reflected in every part. For example, consider the section of coastline photographed from space shown in Fig. 1.2.

You can see that the view from space is similar to the one from 10 miles away, 1 mile away, 1 foot away, 1 inch away, and so on. This is exactly the kind of self-similarity that characterizes fractals.

Let's consider another example of self-similarity. Look at the image in Fig. 1.3 called the *Mandelbrot set*. You will note that there are several globes.

---

[1]For example, the three "Back to the Future" movies.

[2]For example, in George Herbert's *Jacula Prudentum* (1652), referring to the tragic Richard III, "for want of a nail a shoe is lost, for want of a shoe a horse is lost, for want of a horse the rider is lost, [for want of a king, England was lost]." In *Poor Richard's Almanac* (1732), Ben Franklin prefaced the quote with "a little neglect may breed great mischief."

[3]Named after the Latin word *fractus* meaning broken (Mandelbrot, 1982).

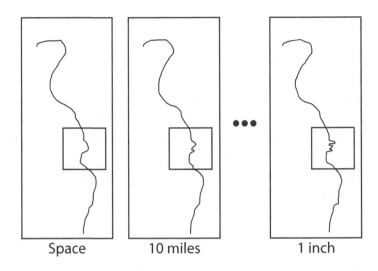

Fig. 1.2.   Increasingly nearer views of a section of coastline.

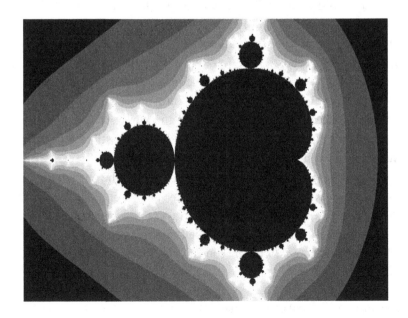

Fig. 1.3.   The Mandelbrot set.

Full-color image:

Look closely at them. Can you see that these globes are just a copy of the larger image? The globes also have a number of "pimples" on them. If you look at them closely too, you'll see that they are also a small reproduction of the larger image. If the image in the figure had an infinite level of detail, you could examine the picture to any magnification and still find a copy of the larger image.

1.4 Other Fractal-like Things

Given self-similarity, other things can appear fractal-like, such as music. Many canons and fugues have a recursive self-similarity. In addition, a school of musical thought called *minimalism* tends to produce sounds that are inherently self-similar.

Some food and baked goods can look fractal-like. For example, many soups, stews and casseroles can appear that way. Pastries made with thin, layered doughs such as Greek baklava or Italian sfogliatelle pastry are fractal-like.

Really, anywhere you can find levels of self-similarity, you can think of fractals. But, in this book, we will mostly focus on fractal images that are created by computer programs.

1.5 How are Fractals Created?

It is unlikely that one could simply "discover" an image that had the property of self-similarity. In the two previous examples, the coastline and the Mandelbrot set, the first was fabricated for illustration (although it could, in theory, exist) and the second was created via a careful mathematical procedure. How do we find such mathematical procedures?

1.5.1 *Dynamical Systems*

Before we answer the question about mathematical procedures, we want to introduce the place where the fractal lives: the realm of dynamical systems. The study of *dynamical systems* is a sub-field of both mathematics and physics that is concerned with the behavior of certain phenomena that depend on time. For example, the weather and tides are time-dependent, so is the movement of the pendulum of a clock or the trajectory of a rocket. A person's height is dependent on time — generally increasing, until their height reaches a maximum and then possibly decreasing with old age. Most chaotic systems are time-dependent. Even aspects of stock markets can be modeled as dynamical systems. Almost anything can be modeled as a dynamical system, though not everything is interesting when it is modeled this way.

Another interpretation is that dynamical systems represent behaviors that occur based on the repeated application of an algorithm. It turns out that any behavior that can be represented as a mathematical function of the time variable, t, can also be represented by a self-referencing function of t, and functions can be represented as program algorithms.

We look at the various aspects of dynamical systems throughout the book and take a deeper dive into them and general systems theory in Chapter 5.

1.5.2 *Algorithms*

An *algorithm* is simply a recipe or set of rules that describe some process. A cookbook contains many algorithms for making food, baking cakes and preparing other goodies. The assembly instructions for a bicycle represent an algorithm.[4] A computer program is simply an encoded form of an algorithm and it is these types of algorithms that you will be seeing in this book.

Algorithms can be presented in many ways: in words, using flow graphs or other diagrams, using pseudo-code (code that is not intended to be run), or in mathematical notation. We will be using combinations of words and mathematical notation throughout the text to describe the algorithms needed to generate fractals. These algorithms involve the application of some function defined on real or complex numbers (to be defined later) or the application of some graphical or geometric procedure. We show how repeated application of either type of algorithm can result in a fractal.

---

[4]The word "algorithm" is derived from the latinized version of "al-Khwarizmi," the birthplace of Mohammad ibn Musa al-Khwarizmi (circa 780–850). He was a Persian mathematician who made significant contributions to algebra and is also associated with the introduction of the decimal numbers in Europe.

Sometimes, an algorithm is called by another name (even though it's still an algorithm) such as rule, method or procedure or even something else. For example, the Sieve of Eratosthenes is an ancient algorithm for finding prime numbers (numbers whose only factors are 1 and itself).

1.5.3 *Attracting and Escaping Points*

Let's begin developing the basic vocabulary needed to describe the algorithms needed for making fractals. Consider a *function* f, which is just a mapping or rule, from the real number line onto itself. We denote this

$$f : \mathbb{R} \longrightarrow \mathbb{R}$$

$$x \longrightarrow f(x)$$

where the symbol $\mathfrak{R}$ stands for the real number line and the arrow, $\longrightarrow$, denotes the fact that the function f is a rule that relates each real number x with another real number $f(x)$. At this point, you might want to get a calculator and follow the discussion along with it.

Consider the function

$$f(x) = x^2$$

If you enter the number 2 and press the "X^2" key, you get 4. Press it again, you get 16, and so on. This procedure is called *function composition*. The composition of $f(x)$ with itself is denoted $f(f(x))$ and it simply means apply the rule f to value x, then apply the rule f again to the result. If we compose the result with the function f again, denoted $f(f(f(x)))$, we say that we have performed another *iteration* of the composition of f. We can continue composing f by itself many times, a procedure we call *function iteration*. For simple functions, iteration is easily performed with a calculator.

Continuing with the example, if you compose the X^2 function enough times, your calculator will probably revert to exponential notation and display something like

3.4028 E 38

which means 3.4028 times the number 1 followed by 38 zeros. Eventually, your calculator will give up and display something like

ERROR

which means that the number obtained was too large for your calculator to hold even using exponential notation. We then say that "the point 2 iterated under the function $f(x) = x^2$ *escaped*," or it tended toward infinity. Functions that tend toward minus infinity at a point under iteration also escape.

For example, the point $x_0 = 2$ iterated under the function $f(x) = -x^2$ will tend toward negative infinity (try it on your calculator). Points that escape under iteration are also sometimes called *repelling* points or are said to be repelled under iteration.

Now, consider iterating any point that is larger than 0 but less than 1, say $x_0 = 0.5$ under the function $f(x) = x^2$. Enter this number into your calculator and press the "X^2" key a few times. You will note that the product gets smaller and smaller. Eventually, the exponential notation will be displayed, but this time it will have a negative exponent, that is, it might look like

2.3283 E -10

which means that the number is 2.3283 times 0.0000000001. This is a very small number. If you keep pressing the "X^2" key, the ERROR indicator may be displayed. This does not mean that the point escaped, rather the number became so close to zero that the calculator could no longer calculate the function X^2 without making an error. In this case, the iterated function tended toward a single point, 0. We would say that 0 is an *attractor* of this function because the iterated function tended toward this point. If you iterate this function with a starting value between -1 and 1 (non-inclusive), the iterated result will always be 0.

Some points act neither as attractors nor as repellers under iteration. Such points are said to be *indifferent*. You can use a scientific calculator or write programs to determine attracting, repelling, or indifferent points for functions (see exercises).

Later on, we will look at the collection of all attracting points for some iterated functions or iterated geometric procedures. When the attracting set of an iterated function or procedures is an infinitely self-similar set, that is, a fractal, then the attracting set is called a *strange attractor*.

1.5.4 *Bifurcation Diagrams*

Let's look at the set of attractors for some functions defined on real numbers.

Suppose we were given the simple polynomial function

$$f(x) = x^2 + c$$

for some real constant c, and we compose the function with itself many times. Let's try this by picking some c, say $c = -1.1$ and set $x = 0$. Using a calculator, we see that

$$f(0) = 0^2 - 1.1 = -1.1$$

Iterating, we find $f(f(0))$, that is,

$$f(f(0)) = f(-1.1) = (-1.1)^2 - 1.1 = .11$$

Now, finding $f(f(f(0)))$, we have

$$f(f(f(0))) = f(.11) = (.11)^2 - 1.1 = -1.0879$$

Applying the rule f to this result again gives

$$f(f(f(f(0)))) = (-1.0879)^2 = .08$$

We could continue this indefinitely, but you should see that the result of the composed functions seems to bounce back and forth between a number somewhere near -1.0 and another number near 0.1. If we compose f itself many times, say 200, and if we do this for a range of values of c, for functions that look like

$$f(x) = x^2 + c$$

a strange and beautiful thing happens.

The image that results from composing this function many times, plotting points after each composition, and doing this for many values of c is called a *bifurcation diagram* and it is shown in Fig. 1.4. The term "bifurcation" is used because the image divides into two distinct bands of points. Like any fractal, it is also self-similar.

Command line

```
cargo run --bin bifurcation
```

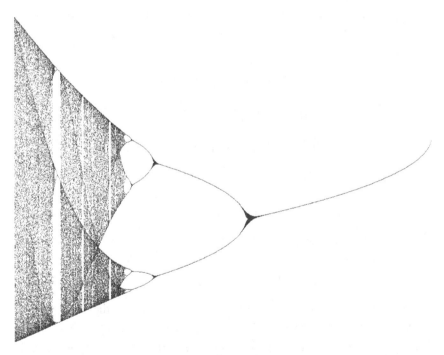

Fig. 1.4.   Bifurcation diagram for $f(x) = x^2 + c$ with $x = 0$ and c swept from -2.0 to 1.0.

Full-color image:

Use the on-screen controls to zoom and pan the fractal. You should see that the bifurcation diagram is self-similar. Also, try changing the range of values over which c is swept.

Note the bands of stability (the "bald spots") where the function only takes on two values instead of infinitely many. The interpretation of these will become clearer later. To create this diagram, we first build a list[5] of equidistant points along the x-axis. The snippet of Rust code is given as follows:

```rust
struct Vertex {
    position: [f32; 2], // array of two elements, x & y
}

// Create n points along the x-axis, where n = screen width
let vertices: Vec<Vertex> = (0..screen_width)
    .map(|x| Vertex {
        position: [
            -1.0 + 2.0 * (x as f32) / (screen_width as f32),
            0.0, // y = 0
        ],
    })
    .collect();
```

Line 6 creates an iterator from 0 to the screen width in pixels (e.g. 800). Line 7 turns each value, x, into a vertex in two-dimensional space. OpenGL, the technology we are using to draw to the screen, uses a coordinate system where each coordinate (x, y, and z) ranges between -1.0 and 1.0. Line 9 is responsible for mapping the values into that coordinate space.

---

[5]Technically, it's a vector, more specifically, Rust's Vec type. In computer science (as well as many programming languages such as Rust and C++), "list" is usually taken to mean a specific data structure called a *linked list*. For our purposes, we will use "list" as interchangeable for "vector." We have no need for linked lists in this book.

Our OpenGL pipeline is configured to treat each vertex as a primitive point. Points are fed one at a time into our geometry shader:

```
1    #version 400
2    #define MAX_VERTICES 256
3
4    layout(points) in;
5    layout(points, max_vertices=MAX_VERTICES) out;
6
7    // some details omitted...
8
9    float map(float x, float in_min, float in_max, float out_min,
     ↪    float out_max) {
10       return (x - in_min) * (out_max - out_min) / (in_max -
     ↪    in_min) + out_min;
11   }
12
13   void main() {
14       float x = gl_in[0].gl_Position[0] / zoom + pan_hor;
15       // Use x to sweep c across its range
16       float c = map(x, -1, 1, c_range[0], c_range[1]);
17
18       float f_x = 0;
19       for (int i = 0; i < 256; i++) {
20           f_x = pow(f_x, 2) + c;
21
22           if (i > 50) { // skip first 50 iterations
23               float f_x_remapped = (f_x - pan_vert) * zoom;
24               gl_Position = gl_in[0].gl_Position + vec4(0.0f,
     ↪    f_x_remapped, 0.0f, 0.0f);
25               EmitVertex();
26               EndPrimitive();
27           }
28       }
29   }
```

For each point, we take the x-coordinate (accessed using `gl_in[0].gl_Position[0]`), scale it by a zoom factor and add a horizontal offset (line 14). We use `map` to convert the x-coordinate (which ranges between -1.0 to 1.0, ignoring zoom and offsets) to a value of c (ranging from, by default, -2.0 to 1.0). In other words, we are sweeping c as we move along the x-axis (line 16).

We then compose the function $f(x) = x^2 + c$ a total of 256 times starting with $x = 0$. The first 50 iterations are ignored to allow the composition to stabilize (lines 18–22). For each subsequent iteration, we plot a point at $(x, f\_x\_remapped)$, where $f\_x\_remapped$ is f_x with the zoom factor and a vertical offset applied (lines 23–26).

Bifurcation diagrams are amazingly simple little fractals that have applications that we will see later. Try experimenting with the bifurcation example program. See what other kind of interesting images you can make by zooming, panning around, and playing with the range of c.

1.6   The Sierpinski Triangle

Another way to generate fractals is by the repeated application of special geometric procedures. Such fractals are called *iterated function systems* (IFS). A nice two-dimensional fractal that can be generated this way is the *Sierpinski triangle*.

Consider a filled triangle. Suppose we remove a section from the middle so that the result is three copies of the original at $1/3$ size, as shown in Fig. 1.5.

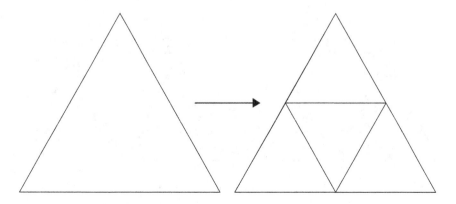

Fig. 1.5.   Rule for creation of a Sierpinski triangle.

If we continue applying this rule to the three triangles, and then the nine resulting triangles and so on, we obtain the fractal shown in Fig. 1.6.

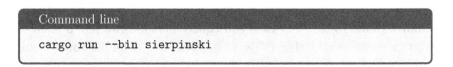

```
cargo run --bin sierpinski
```

Fig. 1.6.   A Sierpinski triangle.

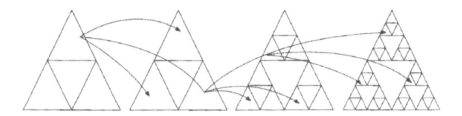

Fig. 1.7. Sierpinski triangle-mapping procedure.

Let's look at how the Sierpinski triangle can be created with a program.

The easiest way to generate such a picture is to generate *random orbits* and look for attracting points. To do this, we select a random starting point and apply the rule to it some fixed number of times. Repeated application of these rules will generate a strange attractor, that is, a fractal.

To make an equilateral Sierpinski triangle, we map the random starting point into one of the three randomly chosen rules corresponding to one of the three triangles in the large triangle with center removed. Figure 1.7 shows how an arbitrary point is mapped into one of the three possible sites.

If you are not already familiar with OpenGL coordinate system, a review of section on "OpenGL coordinate system" in Introduction is suggested at this time. If you do not wish to do so, simply note in the following discussion that OpenGL assigns the coordinate $(-1.0, 1.0)$ to the upper left-hand corner of the screen and $(1.0, -1.0)$ to the lower right-hand corner.

The large triangle has vertices

$$V_1 = (-1.0, -1.0)$$
$$V_2 = (0, 1.0)$$
$$V_3 = (1.0, -1.0)$$

It turns out that if the random point is half way to the outer vertex of one of the three triangles inside the larger triangle, then it is inside one of them. For the three triangles within it, this is also true and so on. To find the half way points to the outer vertices, we use the rules illustrated in Fig. 1.8 and

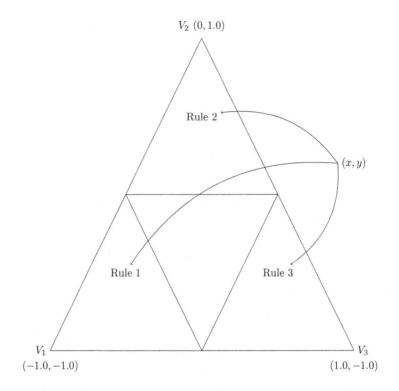

Fig. 1.8.   Vertices for Sierpinski triangle-mapping procedure.

given as follows:

$$(x', y') = \left(\frac{-1.0 + x}{2}, \frac{-1.0 + y}{2} \right) \qquad \text{(Rule 1)}$$

$$(x', y') = \left(\frac{x}{2}, \frac{1.0 + y}{2} \right) \qquad \text{(Rule 2)}$$

$$(x', y') = \left(\frac{1.0 + x}{2}, \frac{-1.0 + y}{2} \right) \qquad \text{(Rule 3)}$$

Each rule finds the half way point between (x, y) and one of the vertices. Rule 1 corresponds to vertex V_1, rule 2 to V_2, and rule 3 to V_3.

If we continually choose one of the three mapping rules and apply them to the coordinate just mapped, we generate points at finer and finer resolution within the Sierpinski triangle.

The code which constructs the fractal is as follows:

```
1   let mut rng = rand::thread_rng();
2   let mut vertices = vec![];
3
4   // Initial starting point
5   let mut x = rng.gen_range(-1.0..=1.0);
6   let mut y = rng.gen_range(-1.0..=1.0);
7
8   for i in 0..200000 {
9       // Generate random integer in range [0, 2]
10      match rng.gen_range(0..=2) {
11          0 => { // rule 1
12              x = (-1.0 + x) / 2.0;
13              y = (-1.0 + y) / 2.0;
14          }
15          1 => { // rule 2
16              x = x / 2.0;
17              y = (1.0 + y) / 2.0;
18          }
19          2 => { // rule 3
20              x = (1.0 + x) / 2.0;
21              y = (-1.0 + y) / 2.0;
22          }
23          _ => unreachable!(),
24      }
25
26      if i >= 1000 { // Skip first 1000 iterations
27          vertices.push(Vertex { position: [x, y] })
28      }
29  }
```

We first choose a random starting x- and y-coordinates (lines 5 and 6). For each iteration, we randomly choose an integer in the range [0, 2] and apply one of the three rules (line 10).[6] Note that we perform 1000 iterations

---

[6]Note that `rng.gen_range(-1.0..=1.0)` returns a floating-point number, i.e. including a decimal place. Possible values include −1.0, −0.3, −0.005, 1.0, etc. If we instead wrote `rng.gen_range(-1..=1)`, then the possible values would be −1, 0, and 1. This is because in Rust, 1.0 is a float, but 1 is an integer.

before plotting to be sure that the fractal has begun attracting. The procedure can be applied to any triangle, equilateral, right, or otherwise.

Another fractal that can be constructed in a similar way is a *Sierpinski*[7] *gasket* or *Sierpinski carpet*. To make a Sierpinski carpet, start with a square, divide it into nine equal-sized squares and remove the middle one. Proceed with the remaining eight squares, repeatedly applying the same procedure. We will see a Sierpinski carpet shortly.

1.7   Iterated Function System Transformations

The geometric rules applied to create the Sierpinski triangles and other fractals can be represented mathematically as a set of operations, including sliding, stretching, and rotating. These types of mathematical operations are called *affine transformations*.

Let's consider how to construct a Sierpinski triangle using affine transformations, given this starter triangle, as shown in Fig. 1.9.

The first step is to scale the triangle to 50% scale, as shown Fig. 1.10.

Then we simply slide (translate) three copies of the scaled triangle southwest, north and southeast, respectively, as shown in Fig. 1.11.

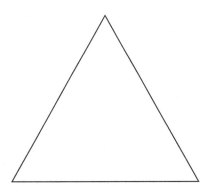

Fig. 1.9.   Starter triangle.

---

[7]Wacław Franciszek Sierpiński (1882–1969) was a Polish mathematician who is known for his contributions to set theory, number theory and topology, a subdiscipline of geometry.

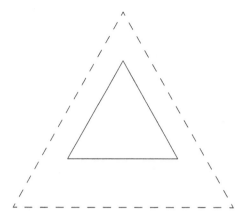

Fig. 1.10. The original triangle (dashed) scaled to 50% scale.

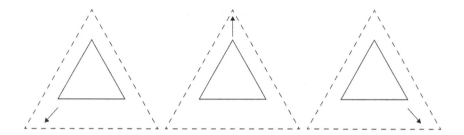

Fig. 1.11. Translations of scaled triangle.

If we superimpose the three scaled and translated triangles, we get the first iteration of a Sierpinski triangle, as shown in Fig. 1.12.

Affine transformations are convenient because they can be coded easily using *matrix* operations.

A matrix consists of rows and columns that hold numbers. If the matrix shown is called "*d*", then the number in the first row first column is denoted $d[1, 1]$, in the second row first column, it is denoted $d[2, 1]$ and so on. In general, the number in row i and column j is denoted $d[i, j]$. Special rules involving the multiplication and addition of the numbers in the matrix simplify the description of affine transformations. We won't review matrix arithmetic here, but you can consult any online or in any book covering the topic of linear algebra.

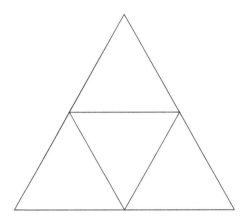

Fig. 1.12.   First iteration of Sierpinski triangle.

Table 1.1.   IFS transformation rule for Sierpinski triangle.

	1	2	3	4	5	6	Probability
1	0.5	0	0	0.5	−0.5	−0.5	0.33
2	0.5	0	0	0.5	0	0.5	0.33
3	0.5	0	0	0.5	0.5	−0.5	0.33

Table 1.1 shows a matrix encoded form of the rules that generate a Sierpinski triangle.

Here, for example, the $d[1, 5]$ position contains number 25. The last column has a special meaning in that it determines the chance or probability that the transformation described in that row will be used. For example, in the Sierpinski triangle, as in sierpinski.rs, the three transformations are equally likely.

Each transformation maps a point (x, y) into a new point (x', y') by handling each coordinate separately. The transformation in row i obtains a new x-coordinate, x', by transforming the given x-coordinate of the point by the mapping

$$x' = d[i, 1]x + d[i, 2]y + d[i, 5]$$

The transformation in row i transforms the y-coordinate by the rule

$$y' = d[i, 3]x + d[i, 4]y + d[i, 6]$$

For example, row 1 specifies[8] that the transformations

$$x' = 0.5x + 0y - 0.5$$
$$y' = 0x + 0.5y - 0.5$$

are to be applied with a probability of 0.33 or one-third of the time.

Hopefully, it is clear how this rule corresponds to the affine transformation illustrated earlier; the x- and y-coordinates are scaled down by 50% and the resulting point is translated down and left (southwest) by 0.5.

Returning to row 1, if we were to apply the transformations to the point $(3, 2)$, we get

$$x' = 0.5 \cdot 3 + 0 \cdot 2 - 0.5 = 1$$
$$y' = 0 \cdot 3 + 0.5 \cdot 2 - 0.5 = 0.5$$

Thus, the transformed point is $(1, 0.5)$. Try applying the transformations to this point for four iterations.

---

[8]Row 1 corresponds to rule 1 of our Sierpinski triangle-mapping procedure (refer to Fig. 1.8). Rewriting the equations a bit may make it easier to see:

$$(x', y') = \left(\frac{-1.0 + x}{2}, \frac{-1.0 + y}{2} \right) \qquad \text{(Rule 1)}$$

$$(x', y') = \left(\frac{1}{2}x - \frac{1}{2}, \frac{1}{2}y - \frac{1}{2} \right) \qquad \text{(Rule 1, tweaked)}$$

Similarly, we can rewrite rules 2 and 3:

$$(x', y') = \left(\frac{1}{2}x, \frac{1}{2}y + \frac{1}{2} \right) \qquad \text{(Rule 2, tweaked)}$$

$$(x', y') = \left(\frac{1}{2}x + \frac{1}{2}, \frac{1}{2}y - \frac{1}{2} \right) \qquad \text{(Rule 3, tweaked)}$$

We take advantage of the simplicity of the matrix form of the IFS in many of our fractal programs. For instance, in program sierpinski-ifs.rs, the code that creates the matrix and samples it repeatedly to generate the Sierpinski triangle is as follows:

```
1   let d = array![
2       [0.5, 0.0, 0.0, 0.5, -0.5, -0.5, 0.33],
3       [0.5, 0.0, 0.0, 0.5, 0.0, 0.5, 0.33],
4       [0.5, 0.0, 0.0, 0.5, 0.5, -0.5, 0.33]
5   ];
6
7   let probs: Vec<f32> = d.slice(s![.., -1]).to_vec();
8   let dist = WeightedIndex::new(probs).unwrap();
9   let mut rng = rand::thread_rng();
10
11  // Initial starting point
12  let mut x: f32 = 0.0;
13  let mut y: f32 = 0.0;
14
15  let mut vertices = vec![];
16  for i in 0..200000 {
17      let r = d.row(dist.sample(&mut rng));
18      x = r[0] * x + r[1] * y + r[4];
19      y = r[2] * x + r[3] * y + r[5];
20
21      if i >= 1000 { // Skip first 1000 iterations
22          vertices.push(Vertex { position: [x, y] })
23      }
24  }
```

```
Command line

cargo run --bin sierpinski-ifs
```

Table 1.2. IFS transformation rule for Sierpinski carpet.

	1	2	3	4	5	6	Probability
1	0.33	0	0	0.33	−0.66	0.66	0.125
2	0.33	0	0	0.33	0.0	0.66	0.125
3	0.33	0	0	0.33	0.66	0.66	0.125
4	0.33	0	0	0.33	−0.66	0.0	0.125
5	0.33	0	0	0.33	0.66	0.0	0.125
6	0.33	0	0	0.33	−0.66	−0.66	0.125
7	0.33	0	0	0.33	0.0	−0.66	0.125
8	0.33	0	0	0.33	0.66	−0.66	0.125

The amazing thing about an IFS is that by simply changing the numbers in the matrix, vastly different fractal images can be generated. In his book, *Fractals Everywhere*, Barnsley (2012) gives the matrix form codes to generate a variety of iterated function system fractals (he calls programs used to generate fractals this way the "Chaos Game").[9]

For example, to make the Sierpinski carpet, simply change the data matrix to that shown in Table 1.2.

This was done in program carpet-ifs.rs, which displays the Sierpinski carpet shown in Fig. 1.13.

Command line

```
cargo run --bin carpet-ifs
```

---

[9]Since we are using a different coordinate system (namely, OpenGL), we've had to adapt the matrices a bit.

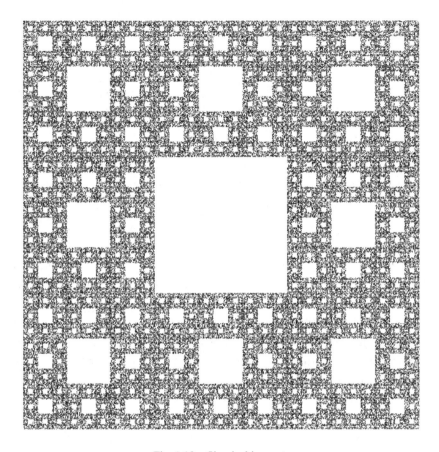

Fig. 1.13. Sierpinski carpet.

In this case, the new data IFS array is as follows:

```
1  let inc = 0.66;
2  let d: Array<f64, Ix2> = array![
3      [0.33, 0.0, 0.0, 0.33, -inc, inc, 0.125],
4      [0.33, 0.0, 0.0, 0.33, 0.0, inc,  0.125],
5      [0.33, 0.0, 0.0, 0.33, inc, inc, 0.125],
6      [0.33, 0.0, 0.0, 0.33, -inc, 0.0, 0.125],
7      [0.33, 0.0, 0.0, 0.33, inc, 0.0, 0.125],
8      [0.33, 0.0, 0.0, 0.33, -inc, -inc, 0.125],
9      [0.33, 0.0, 0.0, 0.33, 0.0, -inc, 0.125],
10     [0.33, 0.0, 0.0, 0.33, inc, -inc, 0.125],
11  ];
```

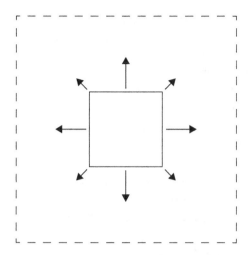

Fig. 1.14. Affine transformations for constructing first iteration of Sierpinski carpet. Original square (dashed) is scaled to 33% scale. Then it is cloned and translated in eight different directions.

As with the Sierpinski triangle, it is helpful to visualize the affine transformations that the matrix represents (Fig. 1.14).

Experimenting with program sierpinski-ifs.rs, carpet-ifs.rs or the other programs using the IFS data algorithm that we will see later produces amazing results. Many of these can be seen in Chapters 3 and 4.

1.8 Recursive Generation of Fractals

Most fractals are generated by applying a certain procedure infinitely (or at least a very large albeit finite number of times). We generated our first fractal, the bifurcation diagram, by repeated iteration of a function. In the last two examples, we iterated random functions by applying repeated geometric procedures. Another way to apply geometric procedures, however, is by coding them so that they are self-referential or *recursive*. Jonathan Swift[10]

---

[10]Jonathan Swift (1667–1745) was an important British essayist, satirist and poet. His most famous work is *Gulliver's Travels*.

captured the spirit of recursion in a poem:

> So, Nat'ralists observe, a Flea
> Hath smaller Fleas that on him prey,
> And these have smaller Fleas to bite'em,
> And so proceed *ad infinitum.*[11]

In mathematics, self-reference usually implies recursion in the sense that a function is defined in terms of itself.

For example, consider the numbers in the famous *Fibonacci*[12] *sequence.* Let $f(0) = 0$ and $f(1) = 1$ be the first two numbers in the sequence. Then the nth number in the sequence $f(n)$ is given by

$$f(n) = f(n-1) + f(n-2)$$

That is, the nth number in the sequence is just the sum of its predecessors. Then the first few numbers in the sequence are

$$0 \ 1 \ 1 \ 2 \ 3 \ 5 \ 8 \dots$$

What would the 23rd number in the sequence denoted $f(23)$ be? Well, you say, it is just $f(22) + f(21)$. But what are these? In fact, you would have to perform a large number of calculations to find $f(23)$ this way. There is a graphical technique using a triangular visualization called Pascal's triangle, but you would need a very large sheet of paper and a lot of time to find these numbers using the technique developed by Pascal.[13] Is there an easier way?

---

[11] *On Poetry: A Rapsody*, 1733.

[12] Leonardo Bonacci (1170–1240), also known as Fibonacci (or "son of Bonacci"), was an Italian mathematician. He wrote one of the most important mathematics texts of the middle ages, *Liber Abaci* or *Book of Calculation*, where he introduced the famous sequence.

[13] French mathematician Blaise Pascal 1623–1662 is known for many achievements in geometry, physics, philosophy and mathematics, including the development of Pascal's triangle. In addition to several theorems named after him, an important programming language, Pascal, also shares his name.

Yes, there is. It is an amazing fact, and part of the allure of mathematics, that you can find an algebraic solution for $f(23)$ or $f(n)$ in general directly, thanks to the Binet[14] formula introduced in 1843.

For the Fibonacci sequence, with $n > 0$,

$$f(n) = \frac{\sqrt{5}}{5}\left(\frac{1+\sqrt{5}}{2}\right)^n - \frac{\sqrt{5}}{5}\left(\frac{1-\sqrt{5}}{2}\right)^n$$

So, if you plug $n = 23$ into the formula, you get

$$f(23) = 28657$$

Just for fun, try to find $f(22)$ and $f(21)$ using this formula, then show that

$$f(23) = f(22) + f(21)$$

Certain programming languages, such as C, C++, Java and Rust, support recursion in the sense that programs and functions of programs may call themselves. For example, consider the Rust program fibo.rs, which finds the nth number in the Fibonacci sequence. It makes use of the self-referential function, `fibo`, given as follows:

```rust
fn fibo(n: i64) -> i64 {
    match n {
        0 => 0,
        1 => 1,
        _ => fibo(n - 1) + fibo(n - 2),
    }
}
```

Note how the function calls itself. You should study the program and try running it with various numbers. We will be seeing other recursive programs later.

---

[14]Jacques Binet (1786–1856) was a French mathematician who made important contributions to physics and mathematics, particularly in number and matrix theories.

Recursion and self-reference can be found in visual art as well. For example, the work of celebrated artist Escher demonstrates an incredible insight into these concepts. For example, in the work *Fish and Scales* (woodcut, 1959), the scales of the fish are themselves fish at many levels, and in Escher's *Circle Limit I* (woodcut, 1958) and *Circle Limit II* (woodcut, 1959), a high degree of self-reference is present. To see these and other beautiful works of Escher, and read his own insightful narrative, see *Escher on Escher: Exploring the Infinite* (Escher, 1989). To learn more about recursion in mathematics, music, art, and life, read *Gödel, Escher, Bach: An Eternal Golden Braid*, the Pulitzer Prize-winning book by Hofstadter (2009).

1.8.1 *The Cantor Set*

To illustrate mathematical recursion again, let's look at the fractal created when we recursively apply the following procedure to a section of the real line:

> remove the middle third of the real line, then remove the middle third of the remaining line segments, and so on.

This procedure, called the *Cantor*[15] *middle third argument*, was introduced in the late 19th century and has a very powerful result. The effect of applying the procedure an infinite number of times is illustrated in Fig. 1.15.

Fig. 1.15.   Construction of the Cantor set.

---

[15]Georg Cantor (1845–1918) was a German mathematician who made important contributions to number theory, logic and other areas. He also invented set theory.

The resulting figure is sometimes called the *Cantor set*, and it is a fractal in one-dimensional *Euclidean space*.[16] If you look at it closely, you should see that there are infinite levels of self-similarity.

When we perform the Cantor procedure an infinite number of times, a very strange thing happens — we never completely eliminate the line. In fact, there will be infinite number of minuscule line segments. Yet, if you strung them all together, their length would be zero.

If you have trouble believing the first part of the previous statement, suppose you were placed at the end of a room that was exactly 20 feet long. You are then asked you to halve your current distance to the other side of the room and to repeat this procedure. Your distance from the opposite wall would then be

$$10 \quad 5 \quad 2.5 \quad 1.25 \quad 0.625\ldots$$

feet and so on. But would you ever reach the other wall? The answer is no — although you can get arbitrarily close to the wall, you can never actually arrive at it using the procedure outlined. The effect is similar when we apply the Cantor procedure, that is, we never completely annihilate the line.[17]

The program cantor.rs applies the Cantor procedure recursively to a line segment on the screen.

Command line

```
cargo run --bin cantor
```

---

[16]Euclidean space is usually thought of as the three-dimensional space in which we live. A computer or TV screen is two dimensional and a number line one dimensional.

[17]There is a paradox here, however. If you rigorously set this up as an equation with limits and solve it for an infinite number of iterations, the answer is that you do reach the wall. This is counterintuitive and contradictory. In fact, this is well known in mathematicians as Zeno's "Achilles Paradox." But so as not to spoil the fun, we'll assume you can't reach the wall this way — who has the time to do it infinitely anyway?

The heart of the code for generating the Cantor set is a recursive procedure called `cantor`:

```
1   fn cantor<L, R>(lines: &mut Vec<Line>, left: L, right: R,
    ↪  depth: u8)
2   where
3       L: Into<Vertex>,
4       R: Into<Vertex>,
5   {
6       let left = left.into();
7       let right = right.into();
8       lines.push(Line(left, right));
9
10      // Keep track of recursion depth
11      if depth >= 6 {
12          return;
13      }
14
15      // Shift subsequent lines down a bit
16      let y = left.y() - 0.05;
17      // Calculate third of line segment
18      let delta = (right.x() - left.x()) / 3.0;
19      // Draw left third
20      cantor(lines, [left.x(), y], [left.x() + delta, y], depth
    ↪  + 1);
21      // Draw right third
22      cantor(lines, [right.x() - delta, y], [right.x(), y],
    ↪  depth + 1);
23  }
```

`Line` is simply a two tuple of vertices. The first vertex is the left endpoint and the second is the right endpoint.

```
1   struct Line(Vertex, Vertex);
2
3   impl Line {
4       fn into_vertices(self) -> impl Iterator<Item = Vertex> {
5           // Return an iterator that yields our two endpoints
6           iter::once(self.0).chain(iter::once(self.1))
7       }
8   }
```

To use `cantor`, we give it a starting line. Each call to `cantor` begins by adding the line formed by the `left` and `right` endpoints to the list. Then we calculate the width of a third of the line and recurse on the left and right thirds. Recursion continues until a hardcoded depth is reached. Afterwards, we use the `Line::into_vertices` method to go from `Vec<Line>` to `Vec<Vertex>`.[18]

```
1   let mut lines = vec![];
2   cantor(&mut lines, [-1.0, 0.95], [1.0, 0.95], 0);
3
4   let vertices: Vec<_> =
    ↪  lines.into_iter().flat_map(Line::into_vertices).collect();
```

That's all there is to it. You should run program cantor.rs yourself, and see that it works.

Repeated application of different geometric rules can generate other fractals. However, recursive geometric rules can generally be performed by an IFS system, which is easier to code. However, finding the IFS codes that are equivalent to the recursive geometric procedure is often quite difficult.

1.9 Fractal Dimension

Suppose that you were given a piece of tin foil with a thickness of exactly zero. We would say that the foil has a Euclidean dimension of two, that is, it is two-dimensional (the Cartesian plane is two-dimensional). Suppose now that you took that piece of foil and crumpled it into a ball. Although the ball now exists in three-dimensional space, it is not quite three-dimensional because it really is not a solid (remember the foil has zero thickness) and cannot be described precisely using Euclidean geometry. We say that the ball has a fractional or *fractal dimension*.

As you might expect, fractal images also have a fractal dimension. Consider, for example, the Cantor set. Normally, a line (with thickness of zero) is said to have a dimension of one (it is one-dimensional). However, we see that

---

[18]This is because OpenGL operates on vertices. In this case, we choose to declare that each pair of vertices represents a line.

the Cantor set is not quite a line, rather a collection of an infinite number of disconnected points. What then is the fractal dimension of the Cantor set?

Consider next the Sierpinski triangle or Sierpinski carpet. Both appear to be two-dimensional, but since they are not filled, their fractal dimension should be somewhat less than two. What then are their fractal dimensions?

There are precise answers to these questions, which are far beyond the scope of this text and most undergraduate mathematics courses. However, using an approximate technique, we can get a feel for the fractal dimension of some of the images described in this text. To do this, however, we need to introduce the concept of logarithms and exponential functions.

First, recall the notation

$$x^y \equiv x \cdot x \cdots x \tag{1.1}$$

that is, we multiply x by itself y times, where y is said to be the *exponent*[19] and x is the base. We say this as "x raised to the power y." For example,

$$4^3 = 4 \cdot 4 \cdot 4 = 64$$

Now, suppose we are given x^y, the number x, and we wanted to find the number y. To do this, we apply a special operation called the base x *logarithm* to x^y. The result will be y and is denoted

$$\log_x(x^y) = y \tag{1.2}$$

From the previous example, we see that

$$\log_4(64) = 3$$

So, how do we use logarithms to find fractal dimension? It turns out that a good approximation of fractal dimension D is

$$D = \log_{10}(\text{number of pieces})/\log_{10}(\text{magnification}) \tag{1.3}$$

which uses the logarithm.

---

[19]We assume that y is a positive integer.

Following is a description of the formula for D in words:

> For a given magnification of a fractal image, we count the number of self-similar pieces, take the logarithm to the base 10 and divide by the base 10 logarithm of the magnification.

For example, consider a solid line. At magnification n, we can divide it into n equal or self-similar pieces of length $1/n$. Thus, the solid line has fractal dimension

$$D = \log_{10}(n)/\log_{10}(n) = 1$$

Now, look at the Cantor set. Since it was produced by removing the middle third from a line, a magnification by three yields two self-similar pieces. Using a calculator, you can then find the fractal dimension to be

$$D = \log_{10}(2)/\log_{10}(3) = 0.63\ldots$$

Next, consider a square on a two-dimensional plane. At magnification n, the square is divided into n^2 self-similar squares, so it has fractal dimension

$$D = \log_{10}(n^2)/\log_{10}(n) = 2$$

To verify this, pick any number $n > 1$ and plug it into the above formula and work it out on a calculator. But for the Sierpinski triangle, any magnification by two yields three self-similar pieces so that it has fractal dimension

$$D = \log_{10}(3)/\log_{10}(2) = 1.58\ldots$$

1.10   How are Fractals and Chaos Related?

There is an intricate and often subtle relationship between fractals and chaos. One way of interpreting their relationship is to note that fractals are generated by detecting attractors and repellers. Thus, they represent, in a sense, a visual representation of chaotic behavior. By plotting points on the real line or Cartesian plane, attractors (stable points) in one color, repellers (chaotic points) in another, black-and-white fractal images can be created. By keeping track of the "speed" at which attractors attract and repellers repel, and plotting bands of these rates in different colors, elegant colored fractals can be generated.

In addition, fractals are chaotic in that they are very sensitive to changes in initial conditions. For example, you will see that changing the function to be iterated even slightly results in a vastly different fractal image output. This sensitive dependence on initial conditions is a theme that unites unstable systems, fractals and chaos.

Another way to see the relationship between fractals and chaos is to study a special mathematical abstraction from dynamical systems called *cellular automata*. It turns out that cellular automata can be stable or chaotic, and many generate fractals. We study cellular automata in Chapter 4.

1.11   Brief History of Chaos, Fractals and Dynamical Systems

It would be very difficult to trace the pedigree of chaos or fractals precisely. To begin with, one would have to study dynamical systems, nonlinear mathematics, functional analysis and so forth. In fact, listing the names of those who have contributed, at least in part, to the theory of dynamical systems is like reading a *Who's Who of Mathematics* (you will see more on all of this in Chapter 5).

But although the early threads of chaos and fractal theory are old, the science itself is very new. Shortly after the First World War (1919), Julia[20] began work on what would later be called attractive cycles of complex functions, but for the next 50 years, most of his work lay dormant.

Then Mandelbrot came along in the early 1980s and showed that natural and human-made phenomenon could be associated with the self-similarity in clearly using fractals (Mandelbrot, 1982). A great deal of work in fractals emerged since that of Mandelbrot, and in years since Mandelbrot's first publications, scientists in diverse fields linked fractals and chaos to their work.

---

[20]Gaston Julia (1892–1978) was a French mathematician born in Algeria whose work in functional analysis laid important foundations for the theory of fractals.

In the 1970s, many scientists such as Devaney,[21] Peitgen[22] and Barnsley[23] popularized and extended the work of Julia and Mandelbrot. For a good presentation on the history of chaos and fractals through the early 1990s, see James Gleick's book, *Chaos: Making a New Science* (Gleick, 2008).

Since the early 1990s, chaos theory and fractals (and their underlying dynamical systems) have appeared regularly in movies that featured chaotic phenomena or had fractal-like visualizations. These include *Jurassic Park* (1993), *The Day After Tomorrow* (2004), *The Matrix* (1999, 2003, 2021), *Dr. Strange* (2016, 2022) and *Jurassic World Dominion* (2022). Certain music has fractal-like qualities (for example, the music of composers Philip Glass and Brian Eno, and the later works of the group, Devo). It is astounding that the ideas of chaos and fractals are no longer only of interest to mathematicians, engineers and scientists — they have become part of our every day life.

1.12   Exercises and Things to Do

Exercise 1.1
Using a scientific calculator, determine an attracting, repelling and indifferent point for the following functions:

(1) $f(x) = x^3 - 1$
(2) $f(x) = -2x(2 - x)$
(3) $f(x) = sin(x)$
(4) $f(x) = 1/x$
(5) $f(x) = \frac{sin(x)}{x}$

---

[21]Robert Devaney, born 1946, is an American mathematician and one of the pioneers in the study of chaos theory and fractals.
[22]Heinz-Otto Peitgen, born in 1945, is a German mathematician who helped popularize chaos and fractals with the general public.
[23]Michael Barnsley, born 1946 is a British mathematician who helped popularize chaos and fractal theory and conducted pioneering work in fractal compression algorithms.

You can also write a program in your favorite language to answer this question.

Exercise 1.2

Repeat Exercise 1.1 by writing a program (or programs) in your favorite computer language. You can expand this program to create fractals for each of the functions.

Exercise 1.3

For the Fibonacci sequence, f, using a program or calculator, find the following:

(1) $f(7)$
(2) $f(13)$
(3) $f(18)$
(4) $f(31)$

Exercise 1.4

Research the Sieve of Eratosthenes and write a program in Rust (or another language) to find the first 100 prime numbers. How long did it take for the program to run? If it was longer than you expected, can you explain why?

Exercise 1.5

Describe another chaotic systems that you encounter in life. Why is it chaotic? What are some of the tipping points in this system?

Exercise 1.6

Listen to "How Now — A Fractal Malestrom" and "Fractal Zoom" by Philip Glass (these can be found on the Web). Describe the fractal nature of these pieces.

Exercise 1.7

Watch any of the movies mentioned in this chapter and write a one-page essay describing how chaos or fractals are used as elements of the film.

Exercise 1.8

Find a recipe for a food or pastry that appears to be fractal-like. See if you can make this food or pastry or, at least, purchase it. Study its fractal qualities and discuss before eating it.

Exercise 1.9

Write a one page biography for one of the following individuals: Fibonacci, Gaston Julia or Benoit Mandelbrot.

Exercise 1.10

Read the book *Chaos: Making a New Science* (Gleick, 2008) and write a one-page summary.

Chapter 2

Foundations of Chaos and Fractal Theory

"From Nature's chain whatever link you strike,
Tenth, or ten thousandth, breaks the chain alike."

— Alexander Pope, *Essay on Man*

In this chapter, we introduce the mathematical background that is needed to generate some of the more breathtaking fractal images. Don't be intimidated if you have never seen this material before. Most of the mathematics involve simple variations on algebraic concepts learned in high-school algebra or trigonometry classes. If you are not comfortable with the mathematics, simply skip the formulas for now, and if you like, come back to them later. However, the formulas do help to unlock some of the mysterious beauty of the fractal images and are worth the effort to master. All of the complex operations described here are incorporated in the code segments in the upcoming discussion. Looking at the code may help you understand the mathematics.

2.1 Complex Numbers and Functions

Consider the mapping signified by the symbol $\sqrt{}$ and generally called the "principal square root." This mapping represents the inverse of the function

$$f(x) = x^2$$

defined on the real line. For positive numbers and 0, this is well defined. However, for negative numbers, it is undefined. For example, what is $\sqrt{-5}$, that is, what number multiplied by itself yields -5?

To get around this problem, Mathematicians[1] have defined the abstract notion of the square root of -1, denoted i, that is,

$$i^2 = -1$$

You should realize that i is only the positive square root of -1 and that a negative square root, say $k = -i$, exists as well, since

$$k^2 = (-1i)^2 = i^2 = -1$$

However, we are only interested in the positive square root.

Note how i takes care of the square root of all negative numbers, since, for example,

$$\sqrt{-4} = \sqrt{-1 \cdot 4} = \sqrt{4} \cdot i = 2i$$

Suppose now that we have a number z of the following format:

$$z = a + bi$$

z is called a *complex number* where a is called the *real part* and b is called the *imaginary part*.For example,

$$4 + 5i, \quad -3.12 + .01i, \quad 0 + 9i, \quad 2.7 + 0i$$

are all complex numbers. *Complex variables* (placeholders for complex numbers) are usually denoted with some variation of the letter z, for example, z_1, z_2, and so on.

2.1.1 *Plotting Complex Numbers*

A complex number can be plotted as a point on the Cartesian plane by letting the real part represent the x-coordinate and the imaginary part represent the y-coordinate. For example, consider the complex numbers, $z_1 = -1 + 2i$,

---

[1]Engineers denote the positive square root of -1 as j. Engineers prefer using j because the variable i generally represents electrical current in engineering equations.

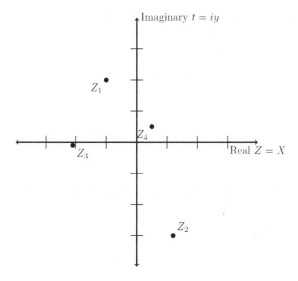

Fig. 2.1. The complex plane and some points on it.

$z_2 = 1.2 - 3i$, $z_3 = -2.1 - 0.1i$, and $z_4 = 0.5 + 0.5i$. These are plotted in on a Cartesian plane, except that the y-axis is labeled as the "iy"-axis, as shown in Fig. 2.1. This map is called the *complex plane* and it is where complex numbers are found.

Functions of complex numbers, which we will see shortly, can also be plotted on the complex plane.

2.1.2 *Arithmetic with Complex Numbers*

Complex numbers can be added, subtracted, multiplied and divided. Addition and subtraction are the easiest to perform — simply add or subtract the respective real and imaginary parts of the numbers. Officially, let $z_1 = a_1 + b_1 i$ and $z_2 = a_2 + b_2 i$ be any complex numbers then for addition

$$z_1 + z_2 = a_1 + a_2 + (b_1 + b_2)i$$

while for subtraction

$$z_1 - z_2 = a_1 - a_2 + (b_1 - b_2)i$$

For example, let $z_1 = 4.1 + 3.1i$ and $z_2 = -1 + 0.1i$, then

$$z_1 + z_2 = 4.1 + -1 + (3.1 + 0.1)i$$
$$= 3.1 + 3.2i$$

and

$$z_1 - z_2 = 4.1 - (-1) + (3.1 - 0.1)i$$
$$= 5.1 + 3.0i$$

The following GLSL functions will perform addition and subtraction of two complex numbers, respectively:

```
vec2 complex_add(vec2 z1, vec2 z2) {
    return vec2(z1.x + z2.x, z1.y + z2.y);
}

vec2 complex_sub(vec2 z1, vec2 z2) {
    return vec2(z1.x - z2.x, z1.y - z2.y);
}
```

GLSL doesn't have a built-in complex number type, so we repurpose vec2, a vector of size 2 where both elements are single-precision floating-point numbers. We treat the x element as the real part and the y element as the imaginary part.

Multiplication and division of complex numbers is just a bit trickier. Consider multiplication first. Let $z_1 = a_1 + b_1 i$ and $z_2 = a_2 + b_2 i$ then by multiplication using the FOIL method (first, outer, inner, and last products) and simplifying[2]

$$z_1 \cdot z_2 = a_1 a_2 + a_1 b_2 i + a_2 b_1 i + b_1 b_2 i^2$$
$$= (a_1 a_2 - b_1 b_2) + (a_1 b_2 + a_2 b_1)i$$

For example, let $z_1 = 4.1 + 3.1i$ and $z_2 = -1 + 0.1i$, then

$$z_1 \cdot z_2 = -4.41 - 2.69i$$

---

[2]In case you are wondering where the $-$ came from, remember that $i^2 = \sqrt{-1}^2 = -1$.

The following GLSL code will perform multiplication of two complex numbers. The input and output to the procedure is in the same manner as for the addition and subtraction procedures:

```
vec2 complex_mult(vec2 z1, vec2 z2) {
    return vec2(z1.x * z2.x - z1.y * z2.y,
                z1.x * z2.y + z1.y * z2.x);
}
```

Next, consider division. Let $z_1 = a_1 + b_1 i$ and $z_2 = a_2 + b_2 i$, then

$$\frac{z_1}{z_2} = \frac{a_1 + b_1 i}{a_2 + b_2 i}$$

We don't know how to do this division directly. To put the ratio in a form that we can handle, multiply numerator and denominator by $a_2 - b_2 i$. This is like multiplying it by 1, which does not change the equality.[3] We then get

$$\frac{(a_1 + b_1 i)}{(a_2 + b_2 i)} \frac{(a_2 - b_2 i)}{(a_2 - b_2 i)} = \frac{(a_1 a_2 + b_1 b_2) + (a_2 b_1 - a_1 b_2)i}{(a_2 a_2 + b_2 b_2) + (a_2 b_2 - a_2 b_2)i}$$

$$= \frac{(a_1 a_2 + b_1 b_2) + (a_2 b_1 - a_1 b_2)i}{a_2^2 + b_2^2}$$

$$= \frac{(a_1 a_2 + b_1 b_2)}{a_2^2 + b_2^2} + \frac{(a_2 b_1 - a_1 b_2)}{a_2^2 + b_2^2} i$$

Now, we have expressed the ratio in terms of the real number division of the real and complex parts.

For example, let $z_1 = 4.1 + 3.1i$ and $z_2 = -1 + 0.1i$, then

$$\frac{z_1}{z_2} = \frac{-3.79 - 3.51i}{(-1)^2 + (0.1)^2}$$

$$= \frac{-3.79}{1.01} - \frac{3.51}{1.01} i$$

$$= -3.752 - 3.475i$$

---

[3] The term $a_2 - b_2 i$ is called the *complex conjugate* of $a_2 + b_2 i$.

The following GLSL procedure will perform division of two complex numbers:

```
vec2 complex_div(vec2 z1, vec2 z2) {
    float denom = z2.x * z2.x + z2.y * z2.y;
    float real = (z1.x * z2.x + z1.y * z2.y) / denom;
    float imag = (z2.x * z1.y - z1.x * z2.y) / denom;
    return vec2(real, imag);
}
```

The input and output to the procedure is in the same manner as for the addition and subtraction procedures.

2.2 Functions of Complex Variables[4]

In order to generate some really interesting looking fractals, we need more powerful functions of both real and complex numbers than just the simple polynomials that we have been working with.

The first of these are two simple functions of real variables, called the *hyperbolic sine* and *hyperbolic cosine*, denoted *cosh* and *sinh*, respectively, and defined as follows[5]:

$$cosh(x) = \frac{e^x + e^{-x}}{2} \tag{2.1}$$

$$sinh(x) = \frac{e^x - e^{-x}}{2} \tag{2.2}$$

Using the hyperbolic cosine and sine, along with the cosine and sine function of real numbers, we define the cosine and sine of a complex number z, denoted $cos(z)$ and $sin(z)$, respectively, as follows:

$$cos(z) = cos(x + iy) = cos(x)cosh(y) - isin(x)sinh(y) \tag{2.3}$$

$$sin(z) = sin(x + iy) = sin(x)cosh(y) - icos(x)sinh(y) \tag{2.4}$$

---

[4]If you are unfamiliar with trigonometry or it's rusty, you may wish to skip this section.

[5]Note that these functions are defined in terms of the special constant e, which is roughly equal to 2.718. e has a special importance to mathematicians, scientists, and engineers, which is similar to that of π and i.

Procedures for these operations in GLSL are shown in the following.

```
vec2 complex_cos(vec2 z) {
    float real = cos(z.x) * cosh(z.y);
    float imag = -sin(z.x) * sinh(z.y);
    return vec2(real, imag);
}

vec2 complex_sin(vec2 z) {
    float real = sin(z.x) * cosh(z.y);
    float imag = cos(z.x) * sinh(z.y);
    return vec2(real, imag);
}
```

Finally, a way of finding the exponential of a complex number is needed. To calculate the exponential of a complex number, use the following equation:

$$e^z = e^{x+iy} = e^x \left(cos(y) + i sin(y)\right) \tag{2.5}$$

To this end, it is interesting to note that Euler's equation[6] relates the exponential to the sine and cosine:

$$e^{ix} = cos(x) + i sin(x) \tag{2.6}$$

A GLSL procedure to find this exponential is as follows:

```
vec2 complex_exp(vec2 z) {
    float real = exp(z.x) * cos(z.y);
    float imag = exp(z.x) * sin(z.y);
    return vec2(real, imag);
}
```

Finally, it is interesting to note that from Euler's equation, it can be shown that the sine and cosine functions of a real number can be defined

---

[6]Note that $e^{i\pi} = 1$, thus harmoniously uniting four important constants.

solely in terms of exponentials, namely,

$$cos(x) = \frac{e^{ix} + e^{-ix}}{2} \qquad (2.7)$$

and

$$sin(x) = \frac{e^{ix} - e^{-ix}}{2i} \qquad (2.8)$$

You might want to try to prove this for practice.

2.3 Generating Fractals by Finding Attractors of Complex Functions

In this section, we look at some beautiful fractals that can be generated by finding the attracting (or escaping) points of iterated complex functions. It is a fascinating characteristic of chaotic systems that vastly different fractals can be generated with a slight alteration of the iterated complex function.

2.3.1 *Julia Sets*

The *Julia set* of a complex function $f(z)$ is the boundary of the set of points that escape — points in the Julia set do not themselves escape, but points arbitrarily close by do. We could not possibly determine these points without infinite computing power. Instead, we will find the points that themselves escape, and assume that the points that don't escape are arbitrarily close by.

There are three basic techniques for finding Julia sets of complex functions. The first is by computing escaping orbits. The second is called the inverse iteration method (IIM) and the third is called the boundary scanning method (BSM). The two latter ones are superior to the first, but the first is easier to code and understand, so we will use it throughout the text.

To find escaping orbits, we iterate the function $f(z)$ at each point on a portion of the complex plane centered at $(0, 0)$. We iterate the function until either the point attracts or escapes (indifferent points are treated as escaping).

However, finding escaping and attracting points for complex-valued functions is a little more difficult than for real-valued functions. We can't test complex-valued functions to see if they are less than infinity (and greater than minus infinity) because they have an imaginary part. Instead, we need to find the modulus of the complex function at each iteration and test that.

The *modulus* of a complex number z is equal to the square root of the sum of the squares of its real and imaginary parts (remember the Pythagorean theorem?). That is, if $z = a + ib$, then its modulus, denoted $| z |$, is

$$| z |= \sqrt{a^2 + b^2} \tag{2.9}$$

For example, let $z = 3 + 4i$, then

$$| z |= \sqrt{3^2 + 4^2} = \sqrt{25} = 5$$

Now, we can define attraction and repulsion of complex-valued functions. A function $f(z)$ iterated at point z_0 attracts if the square of its modulus (i.e. the sum of the squares of its real and complex parts) at any point in the iteration is less than some threshold, which we call the *attractor sensitivity*. The threshold is generally set to be much less than 1. In some cases, minor variations in the attractor sensitivity can result in wild variations in the image produced.

If a function is iterated at a point, and after a certain number of iterations it has not attracted, or if its modulus exceeds some number, then the point has escaped. The number of times that a function is iterated before we decide that it has escaped depends on a couple of factors. First, if the modulus of the iterated function is less than the sensitivity, then the point attracts. If the modulus is less than 100, and the number of iterations is less than the maximum, we will continue iterating. Finally, if the modulus of the iterated function reaches or exceeds 100, the point is considered as escaping.[7]

The maximum number of iterations is controlled by the number of colors we want to use to render the fractal.[8] We find that around 100 is a sweet spot.

---

[7]These numbers are entirely arbitrary.

[8]Tying number of colors to iterations has unintended consequences that we'll explore in Section 2.3.2.

Let's take a look at our program to generate Julia sets. The fractal is mostly implemented as a fragment shader:

```
1   void main() {
2       vec2 z = vec2(
3           xMin + (xMax - xMin) * (gl_FragCoord.x / width),
4           yMin + (yMax - yMin) * (gl_FragCoord.y / height));
5
6       const float attract = 0.0001;
7
8       color = vec4(1, 1, 1, 1);
9
10      for (uint i = 0u; i < maxColors * 2u; i++) {
11          // Apply function
12          z = F(z);
13          float mag = length(z);
14          if (mag < attract) {
15              // Point is an attractor
16              break;
17          } else if (mag >= 100) {
18              // Point escaped
19              vec3 s = Colorize(float(i/2.0) /
                    ↪ float(maxColors));
20              color = vec4(s.xyz, 1);
21              break;
22          }
23      }
24  }
```

The first line sets z to the pixel position, scaled and shifted appropriately. By default, xMin & yMin are -2.0 and xMax & yMax are 2.0. This corresponds to a complex plane where the real and imaginary axes range from -2.0 to 2.0 (see Fig. 2.2).

Next, we hardcode the attractor sensitivity. In this case, an iterated function $f(z)$ is assumed to attract at a point if after a suitable number of iterations (maxColors * 2), the square of the modulus (length(z)) is less than 0.0001. You can play with the attractor sensitivity and zoom/pan to obtain different effects.

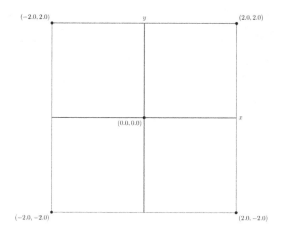

Fig. 2.2. The plotting window for Julia and Mandelbrot programs.

An interesting aspect of this program that we haven't seen yet is the usage of a GLSL feature called **subroutines**. Subroutines are special functions in which the **declaration** (parameter type and return type) is separate from the **definition(s)** (function body). More than one definition can be provided for a given subroutine declaration. At runtime, a subroutine uniform variable is used to switch between the actual function that is used. In practical terms, subroutines provide a mechanism that lets us swap out the definition of $f(x)$ at runtime so we can display different Julia sets.[9]

In the code snippet above, $f(x)$ is called on line 12 by way of invoking the subroutine F. F is declared as follows:

```
subroutine vec2 f_t(vec2 z);
```

The way to read this is "f_t is a subroutine type taking a vec2 and returning a vec2."

```
Command line

   cargo run --bin julia_simple cos
```

The first Julia set we will draw is for the complex cosine function, $f(z) = cos(z)$. It is pictured in Fig. 2.3. Its definition looks like this:

---

[9]We also make extensive use of subroutines to switch between different coloring methods. `Colorize` is a subroutine call that takes a `float` (0.0 to 1.0) and returns a `vec3` (RGB). The default implementation calls `ColorMap` which is another subroutine that uses hardcoded

Fig. 2.3.   Julia set for $f(z) = cos(z)$.

Full-color image:

```
subroutine(f_t)
vec2 FCos(vec2 z) {
    return complex_cos(z);
}
```

Drawing other Julia sets is just a matter of providing additional definitions for F and choosing them via uniform variable. For example, Fig. 2.4 shows what is known as Douady's rabbit.[10] It is generated by finding the

---

color palettes. These indirections allow us to reuse a great deal of code to display a wide variety of fractals.

[10] Adrien Douady (1935–2006) was a French mathematician who built on the work of Julia, making significant contributions to analytic geometry and dynamical systems.

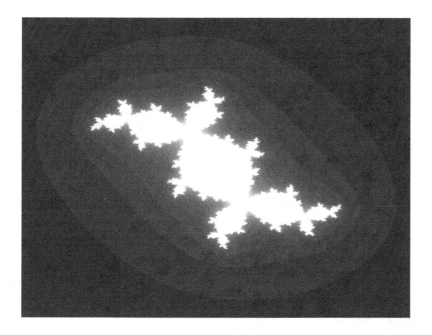

Fig. 2.4. Douady's rabbit.

Full-color image:

Julia set of the function

$$f(z) = z^2 + -0.122 + 0.745i$$

```
Command line

cargo run --bin julia_simple rabbit
```

Its definition is as follows:
```
subroutine(f_t)
vec2 FRabbit(vec2 z) {
    z = complex_mult(z, z);
    z = complex_add(z, vec2(-0.122,0.745));
    return z;
}
```

2.3.2 *The Histogram Coloring Algorithm*

There is an issue with our simple method for coloring the Julia sets. To understand why, let's observe what happens when we generate Douady's rabbit with increasingly higher iteration counts. We will be using the Inferno color scheme, as shown in Fig. 2.5.

We'll start with an iteration count of 10 (Fig. 2.6). For our Douady rabbit Julia set (as configured), the minimum iteration count for points that escape is 3 and the maximum is 10. Figure 2.7 illustrates the portion of the color scheme that is utilized.

Let's crank the iterations up to 50. Figure 2.8 shows the result. The image as a whole has gotten darker. This is because the minimum escape iteration is still 3, but since the color scheme has been subdivided into more pieces, the corresponding color falls farther left on the spectrum.

Based on Fig. 2.9, orange and yellow *should* still be visible in the image. Indeed they are, but you need to zoom in to see them because they correspond to higher iterations.

At 200 iterations, the image (Fig. 2.10) is darker still. Blue is the only easily discernible color, with perhaps a tinge of purple if you look closely. In fact, even if you zoom in, you will not see any other colors. The maximum escape iteration is 75 for this particular fractal. So, only the first $\frac{75}{200} = \frac{3}{8}$ths of the color spectrum is used. The situation is even more grim at 800 iterations where the result is nearly monochromatic (Fig. 2.11).

What we need is a method of choosing colors that is independent of maximum iterations. This is where the histogram coloring algorithm comes in. As a sneak preview, check out Fig. 2.12 — it is the result of

Fig. 2.5.   Inferno color scheme.

Full-color image:

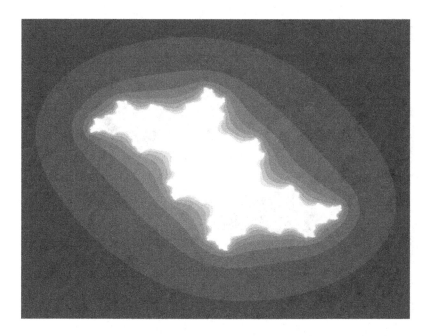

Fig. 2.6.   Douady's rabbit created using 10 iterations.

Full-color image: ▣▒▒▒.

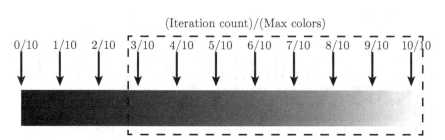

Fig. 2.7.   Inferno color scheme utilization (dashed box) for Douady's rabbit with 10 iterations. Note how the darker colors on the left are not used because the escape iteration count starts at 3.

Full-color image: ▣▒▒▒.

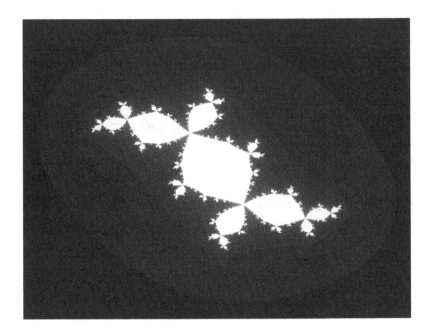

Fig. 2.8. Douady's rabbit created using 50 iterations.

Full-color image:

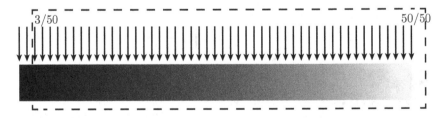

Fig. 2.9. Inferno color scheme utilization (dashed box) for Douady's rabbit with 50 iterations.

Full-color image:

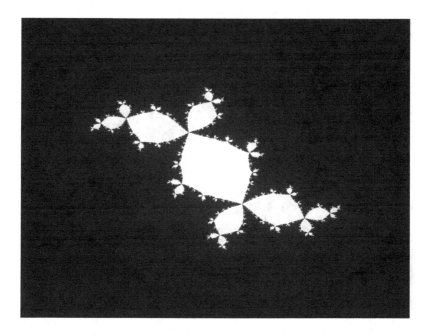

Fig. 2.10. Douady's rabbit created using 200 iterations.

Full-color image:

drawing Douady's rabbit using 800 iterations with histogram coloring. Note how despite the high iteration count, the full gamut of colors is visible.

```
Command line
cargo run --bin mandel_julia rabbit
```

The code for implementing histogram coloring is too complex to fully discuss in this book. Instead, we will just discuss the essential ideas.

The basic approach is to compute the fractal twice: the first time, we find whether each point escapes or not. If a point does escape, we record the number of iterations it took. Once all points have been processed, we create a histogram of escape iteration counts. This gives the distribution of escape iteration counts across the image.

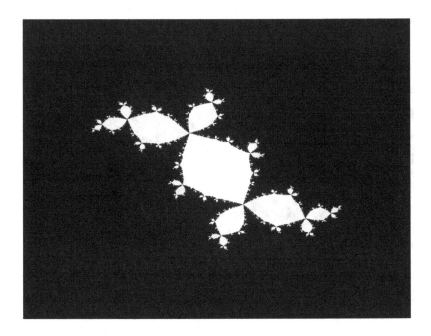

Fig. 2.11. Douady's rabbit created using 800 iterations.

Full-color image:

Next, we consult the histogram to figure out how to map escape iteration counts to positions along the color map such that the full range of colors will be used in the final image. Specifically, we take the octiles (8-quantiles) of the histogram. The zeroth octile is just the minimum iteration count[11] — it is mapped to the leftmost color on the color map. The first octile is the iteration count such that approximately $\frac{1}{8}$th of all the iteration counts are less than this value — it is mapped to the color which is $\frac{1}{8}$th along the color map, and so on. For iteration counts that fall somewhere between the octiles, we linearly interpolate between the corresponding colors.

Finally, we can compute the fractal a second time, using the coloring scheme derived above. This gives us our beautiful (and uniformly)

---

[11]This is 3 in our Douady's rabbit example.

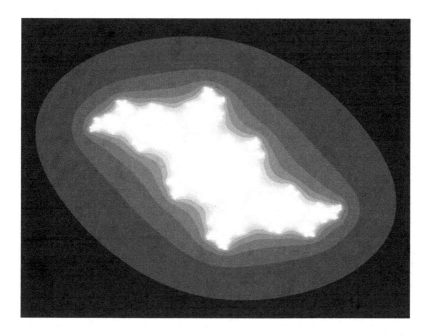

Fig. 2.12. Douady's rabbit created using 800 iterations with histogram coloring.

Full-color image:

colored images. Note that from this point onward, all Julia sets (and related fractals) in the book will use histogram coloring for more pleasing images.[12]

2.3.3 *More Julia Sets*

The "Siegel[13] disk," shown in Fig. 2.13, is a Julia set generated by iterating the function

$$f(z) = z^2 + -0.390540 - 0.58679i$$

---

[12]While histogram coloring is certainly not simple, it is far from the most sophisticated algorithm that has been devised. Another interesting method you should research is continuous coloring which avoids the banding.

[13]Carl Ludwig Siegel (1896–1981) was a German mathematician specializing in number theory.

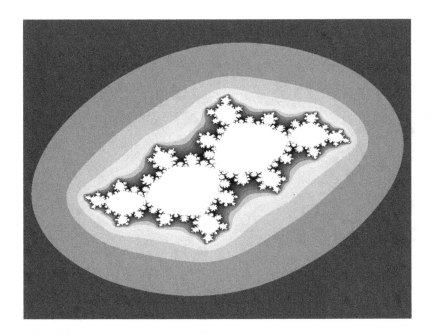

Fig. 2.13.   A Siegel disk.

Full-color image:

```
Command line

cargo run --bin mandel_julia siegel
```

The definition for the Siegel disk is as follows:

```
subroutine(f_t)
vec2 FSiegel(vec2 z) {
    z = complex_mult(z, z);
    z = complex_add(z, vec2(-0.390540,-0.58679));
    return z;
}
```

Next, the dragon-like image shown in Fig. 2.14 can be created by finding the Julia set of t

$$f(z) = z^2 + 0.360284 + 0.100376i$$

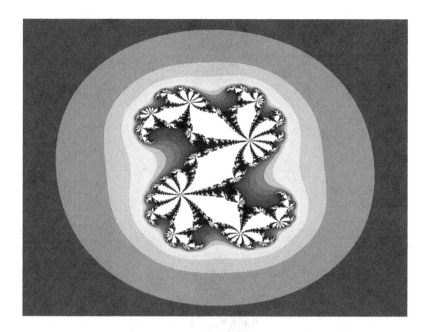

Fig. 2.14. A dragon.

Full-color image:

```
Command line

cargo run --bin mandel_julia dragon
```

The subroutine definition for the dragon is as follows:

```
subroutine(f_t)
vec2 FDragon(vec2 z) {
    z = complex_mult(z, z);
    z = complex_add(z, vec2(0.360284, 0.100376));
    return z;
}
```

Note how just a slight change in the complex constant being added to z^2 dramatically changes the image.

Fig. 2.15. The Julia set of sin(z).

Full-color image:

Finally, let's look at the Julia set of $f(z) = sin(z)$. Its image looks like Christmas ornaments and is shown in Fig. 2.15.

```
Command line

cargo run --bin mandel_julia sin --color-scheme magma
```

Its subroutine definition is simply as follows:

```
subroutine(f_t)
vec2 FSinh(vec2 z) {
    return complex_sin(z);
}
```

Incidentally, for fun, you should try modifying the Julia set code to find the Julia sets of the following functions:

(1) $f(z) = \pi i e^z$
(2) $f(z) = (1 + 0.01i) \sin z$
(3) $f(z) = 2.965 \cos z$

Be sure to use the complex GLSL procedures previously introduced.

2.3.4 *The Mandelbrot Set*

A different kind of fractal that can be generated by finding escaping and attracting points of complex function is the Mandelbrot set. A *Mandelbrot set* is the set of complex constants c_i for which the orbits of the function

$$f(z) = g(z) + c_i \qquad (2.10)$$

evaluated at the initial condition of $z_0 = 0$ do not escape. You may have noted that the Mandelbrot set is somewhat similar to a Julia set, but it is not exactly a graph in the complex plane, rather it is a graph of the parameter space determined by the c_i, where the real part of c_i is plotted on the x-axis and the imaginary part plotted on the y-axis.

Normally, the set contains the points whose orbits do not escape for the function

$$f(z) = z^2 + c_i \qquad (2.11)$$

In this case, $g(z) = z^2$ in equation (2.10). However, the "Mandelbrot set," which is named after its discoverer, Benoit Mandelbrot,[14] can be found for other functions of z.

To write a program that generates Mandelbrot sets, you only have to modify the GLSL code slightly. While with Julia sets we sweep the value of z over some range, here we fix $z = 0$ and sweep the complex constant c. We then check for attracting points, and color escaping points in a similar manner.

---

[14]Benoit Mandelbrot (1924–2010) was a Polish-French-American mathematician and a founder and pioneer in the science of fractals. An IBM scientist for most of his career, he was the first to apply computers to the study of fractals and fractal-like structures.

The pertinent code is as follows:

```
1   void main() {
2       vec2 c = vec2(
3           xMin + (xMax - xMin) * (gl_FragCoord.x / width),
4           yMin + (yMax - yMin) * (gl_FragCoord.y / height));
5
6       uint i = 0u;
7       float mag = 0;
8       const float escape = 4.0;
9       vec2 z = vec2(0, 0);
10
11      while (i++ < 30 && mag < escape) {
12          z = complex_square(z) + c;
13          mag = length(z);
14      }
15
16      if (mag < escape) {
17          color = vec4(0, 0, 1, 1);
18      } else {
19          color = vec4(1, 1, 1, 1);
20      }
21  }
```

Note how we test the square of the modulus, variable `mag`, to see if it is less than the escape threshold. If it is, then the point attracts, and we color the pixel blue at the point. Otherwise, we keep iterating the function up to 30 times.

The result, called the *filled Mandelbrot set* (it uses only one color), is shown in Fig. 2.16.

If, however, we color escaping points in terms of the number of iterations it takes them to escape, and do not color attracting points, then we get the beautiful and well-known image of Fig. 2.17, generally called the "Mandelbrot set."

```
Command line

    cargo run --bin mandelbrot_simple
```

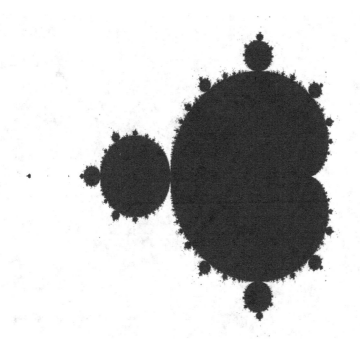

Fig. 2.16. A filled Mandelbrot set.

The pertinent code is as follows:

```
1   while (i++ < maxColors && mag < escape) {
2       z = complex_square(z) + c;
3       mag = length(z);
4   }
5
6   if (mag < escape) {
7       color = vec4(0, 0, 0, 1);
8   } else {
9       vec3 s = Colorize(float(i) / float(maxColors));
10      color = vec4(s.xyz, 1);
11  }
```

Note how we output the pixel color in terms of the number of iterations. As with the Julia set, we can leverage the histogram coloring algorithm (as in Fig. 2.18) to give better results.

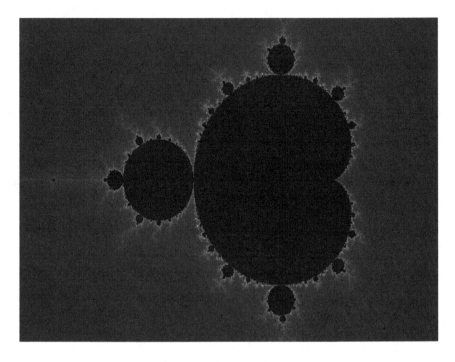

Fig. 2.17. The Mandelbrot set.

Full-color image:

```
Command line
cargo run --bin mandel_julia -- -m --color-scheme magma
```

2.3.5 *A Note on the Images*

By making tiny adjustments to the initial conditions in the Mandelbrot and Julia sets, we can generate an amazing variety of different types of fractals.

We should note here that because these dynamical systems are so sensitive to minor variations in initial conditions, the accuracy of the computer, number of iterations used to determine escape and so forth affect the appearance of the final image. For example, some of the images that we generated

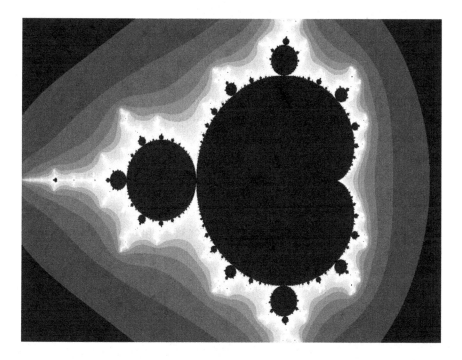

Fig. 2.18.   The Mandelbrot set using the histogram coloring algorithm.

Full-color image:

here, and which you may have seen elsewhere, may differ slightly in levels of detail. The details may be different, but the overall morphology or shape is the same. Finally, the choice of colors and their assignment is arbitrary.

2.3.6   *The Inverse Iteration and Boundary Scanning Methods*

We have used and will be using the method of finding escaping orbits to generate Julia and Mandelbrot sets throughout the text. However, as a point of interest, we should briefly discuss the inverse iteration method (IIM) and the boundary scanning method (BSM).

The IIM essentially takes advantage of the fact that if

$$f(z) = z^2 + c = w$$

then the *inverse*[15] of the function, denoted $f^{-1}(z)$, is

$$f^{-1}(z) = \pm\sqrt{w - c}$$

The orbits are then found by iterating randomly on $+f^{-1}(z)$ and $-f^{-1}(z)$. In other words, IIM takes advantage of symmetry to halve the number of calculations.

The BSM works by sliding a box or window across the region of the complex plane to be plotted. The points at the corners are then tested for attraction. If all four points attract to the same point, then the center of the box is an attractor, otherwise it repels. The technique is a three-dimensional analogy of testing if a car on a roller coaster is at stable or unstable equilibrium. If it is the former, then the cup-like region formed is said to be a *basin of attraction* (see Fig. 2.19).

BSM is truer to the definition of the Julia set and so often generates crisper fractal images. However, it takes far more computations than our simple search for attracting points.

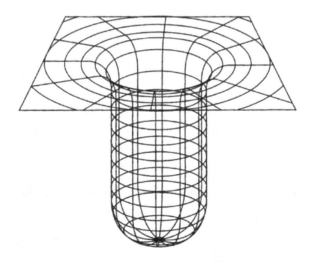

Fig. 2.19. Basin of attraction used in BSM.

---

[15]For a real function f, with inverse f^{-1}, then $f^{-1}(f(x)) = x$.

2.4 Three-Dimensional Fractals

Fractals in three dimensions can be generated, but instead of plotting points on a plane, we need a cube or three-dimensional space. For this purpose, complex numbers are not sufficient, instead more sophisticated mathematical tools called *quaternions* are needed. These are in fact hyper-complex numbers, or in essence, a pair of complex numbers with extended properties.

Quaternions were invented by Hamilton[16] for three-dimensional rotational kinematics (the study of motion) and are quite important for 3D graphics and in the generation of three-dimensional fractals.

A quaternion q is written as

$$q = w + ix + jy + kz \tag{2.12}$$

The numbers i, j, and k are all the positive square root of -1, that is,

$$i^2 = j^2 = k^2 = -1$$

You should note that if the j and k terms are dropped, this is just a complex number. In three-dimensional fractals, if we set $w = 0$, the x, y and z terms are used to select the x, y and z coordinates of a pixel in three-dimensional space.

The manipulation of quaternions is much more complicated than for complex numbers because quaternions and their associated operations form what is termed a *non-commutative algebra*. What this means is very simple. While

$$x \cdot y = y \cdot x$$

for any two real or complex numbers, with quaternions this property does not hold.

---

[16]Sir William Rowan Hamilton (1805–1865) was an Irish mathematician who, while best known for his invention of quaternions in 1843, made many other contributions to optics, classical mechanics and abstract algebra. His refinement of Newton's classical mechanics, which unified the theories of mechanics, optics and mathematics, is known as "Hamiltonian mechanics."

For example, for quaternions, the following is true:

$$i \cdot j = k$$

but

$$j \cdot i = -k$$

In general, quaternions satisfy the multiplicative rules:

$$i^2 = j^2 = k^2 = -1$$
$$i \cdot j = -j \cdot i = k$$
$$j \cdot k = -k \cdot j = i$$
$$k \cdot i = -i \cdot k = j$$

You may have heard of "Maxwell's Equations" — four vector equations that define the fundamental relationships between electricity and magnetism that are among the most important formulae in all physical sciences. The original formulation by Maxwell[17] was in 20 quaternion equations with 20 variables. These were simplified to their modern form (4 equations in 6 variables) by Heaviside.[18]

A further study of quaternions is beyond the scope of this book, but it is interesting to note that three-dimensional fractal-like images can appear in certain applications, for example, in video games, autonomous vehicle displays and robot navigational spaces.

2.5 Wavelets

Wavelets (small waves) refer to self-similar functions generated from one basic function called the *mother wavelet*. The self-similar wavelets (also known as *daughter wavelets*) are generated by time shifting and scaling the

---

[17]James Clerk Maxwell (1831–1879) was a very important Scottish mathematician and scientist whose namesake equations are part of the foundation of electrical engineering.
[18]Oliver Heaviside (1850–1925) was self-taught, yet made many very important contributions to engineering and mathematics.

mother wavelet.[19] Wavelets, which began to be explored in the 1980s, are used in signal analysis, modeling of certain phenomena, medical imaging, image compression and many other areas (Gargour, 2009).

There are several commonly used mother wavelet functions, but let's look at just one for the purposes of illustration. Consider this "Mexican hat" function given by the complicated looking formula:

$$f(t) = \frac{2}{\sqrt{3\sigma}\,\pi^{1/4}}\left(1 - \left(\frac{t}{\sigma}\right)^2\right)e^{-\frac{t^2}{2\sigma^2}}$$

where σ is a parameter that can be adjusted to scale the wavelet.

A plot of this function is shown in Fig. 2.20 (you can see why it is called the Mexican hat function — it looks like a sombrero.).

By tuning the parameter σ, you can generate an infinite number of self-similar wavelet forms. Just a few of these are shown in Fig. 2.21.

Now, scaling and reassembling a number of these wavelets together, we can approximate an electrocardiogram (EKG) image shown in Fig. 2.22.

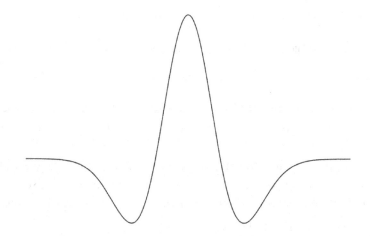

Fig. 2.20.   Mexican hat function mother wavelet.

---

[19]There is a much more rigorous definition of wavelets, but we'll omit this for simplicity of presentation.

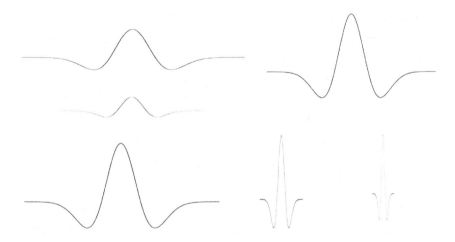

Fig. 2.21.   Some daughter wavelets of the Mexican hat function.

Fig. 2.22.   Simulated electrocardiogram image produced by Mexican hat wavelets.

An electrocardiogram (ECG or EKG) image represents the voltage activity of the heart versus time, and you may have seen it depicted on television show or movies.

So, now if we wish to store the image in Fig. 2.22, we don't have to store all of it — just a copy of the mother wavelet and the rules of reassembly of the many self-similar copies of the mother wavelet. This approach leads to a great way to compress images.

There is a strong connection between wavelets and fractals. Both exhibit self-similarity at different resolutions. Scaling properties exist in both wavelets and fractals. Both can be used in all kinds of image and data analysis, recognition and compression applications.

2.6 Exercises and Things to Do

Exercise 2.1

For $z_1 = 3.4 - 1i$, $z_2 = -7.1 + 3.2i$ and $z_3 = -0.5 - 0.3i$, find the following:

(1) $z_1 + z_2$
(2) $z_2 - z_3$
(3) z_1^2
(4) $z_2 * z_3 - z_1$
(5) z_1/z_3
(6) $z_2^2 + z_3^2/z_1^3$

You can do these calculations by hand using a scientific calculator or write programs to do them. If you choose Rust, consider using the num-complex crate.

Exercise 2.2

For $z_1 = 2.1 - 3.2i$, $z_2 = -5.3 - 0.4i$ and $z_3 = -1.7 - 2.6i$, find the following:

(1) $cosh(z_1)$
(2) $sinh(z_2)$
(3) $sinh(z_2/z_3)/cosh(z_1 - z_3)$
(4) $cos(z_3 - z_)$
(5) $sin(z_3/z_2)$
(6) $cosh(z_1) * cos(z_1) * sinh(z_2) * sin(z_2)$

You can do these calculations by hand using a scientific calculator or write programs to do them. If you choose Rust, consider using the num-complex crate.

Exercise 2.3

Our Mandelbrot programs (mandelbrot_simple.rs and mandel_julia.rs) so far have used the function, $f(z) = z^2 + c_i$. Try modifying either of those programs to plot other functions of z. For example, try $f(z) = z^3 + c_i$

or $f(z) = z^4 + c_i$. Research the "Multibrot set" for more inspiration and analysis.

Exercise 2.4

Investigate other coloring methods for Mandelbrot and Julia sets. A good place to start is https://en.wikipedia.org/wiki/Plotting_algorithms_for_the_Mandelbrot_set#Coloring_algorithms. Try modifying mandel_julia to implement HSV or LCH coloring. You'll need to propagate the highest achieved iteration count from the histogram to the fragment shader (you won't be able to use mandelbrot_simple or julia_simple since those methods don't calculate that information).

Exercise 2.5

Investigate the following mother wavelets:

- Haar
- Gaussian
- Daubechies
- Morlet
- Symlet
- Shannon

Draw the mother wavelet and discuss possible applications for each.

Exercise 2.6

Write a one-page biography of one or more of the following:

- Sir William Rowan Hamilton
- James Clerk Maxwell
- Oliver Heaviside

Chapter 3

Chaos and Fractals in Nature

"And Chaos, ancestors of Nature, hold
Eternal anarchy, amidst the noise."

— John Milton, *Paradise Lost*

In this chapter, we look at natural phenomenon that are chaotic in nature. We also analyze how we can model natural phenomenon in terms of fractals. We will be especially interested in writing programs that can simulate natural beauty.

3.1 Population Dynamics

The chaos of nature can be seen in the study of population dynamics, particularly the relationship between predator and prey populations. Although the models used are necessarily simplistic, they provide significant insight into the interrelationship between animals in a small part of the food chain.

As an example, suppose an ecologist is studying the population of caribou on an island in Canada. They estimate the population of a species in a defined area using the "capture–recapture" method invented by Laplace[1] in 1783. The method involves capturing a number of animals, then tagging

---

[1]Pierre-Simon Laplace (1749–1827) was a French mathematician whose contributions to mathematics, physics and engineering are significant and enduring. His dynamical theories of ocean tides, planetary motion, acoustics and more are most closely associated with the topics of this book. But his mathematical theories and techniques, most notably the "Laplace Transform," are used by engineers every day.

them and releasing them to reintegrate into the population. After a period of time, another group of the animal is captured. The ratio of marked to unmarked animals is a good estimator of the actual population.

The ecologist finds that the population is unstable because of crowding, disease and lack of food. The ecologist proposes to introduce some wolves on the island to help stabilize the population.

Let's model this system and see why it is highly unstable. Let caribou(t), wolf(t) be the number of caribou and wolves in month t, respectively, and caribou$_b$ be the population increase of caribou due to births. If there were unlimited resources of food, space, and so on, then the excess of the birth rate over the death rate for the caribou is positive. In the absence of predators, then the population of caribou grows at a rate of

$$\text{growth}(t) = \text{caribou}_b \cdot \text{caribou}(t) \tag{3.1}$$

To simplify the model, we ignore some of the other ways that the caribou population could increase. Other ways could be in migration (caribou can swim), escapees from captivity, or human initiatives to increase the population by transporting wolves from other locations.

The death rate of caribou from wolves depends on the number of encounters between wolves and caribou, and the success of a kill. Let's call that factor K. K is proportional to the numbers of caribou and wolves, and should be a number much smaller than one. So, the death rate for caribou can be modeled as

$$\text{death}(t) = K \cdot \text{caribou}(t) \cdot \text{wolf}(t) \tag{3.2}$$

Then the equation describing the caribou population is

$$\text{caribou}(t + 1) = \text{caribou}(t) + \text{growth}(t) - \text{death}(t)$$

or

$$\text{caribou}(t+1) = \text{caribou}(t) + \text{caribou}_b \cdot \text{caribou}(t) - K \cdot \text{caribou}(t) \cdot \text{wolf}(t)$$

Note that because both K is less than one, the population for caribou in subsequent months may be non-integer, for example, you could have 31.6 caribou. We could handle this by rounding or truncation (ignoring the numbers after the decimal), but since the model is already an approximation, it is fine to carry forward a fractional number for the population of caribou and wolves.

Now, let's look at the wolf population dynamics. To simplify the model, assume that the death of each caribou result in the birth of one wolf. Further, we assume that this is the only means by which the wolf population can grow.[2] However, it is subject to a death rate of $wolf_d$. Thus, the wolf population is determined by

$$wolf(t+1) = wolf(t) + K \cdot caribou(t) \cdot wolf(t) - wolf_d \cdot wolf(t) \quad (3.3)$$

Again, this could result in a non-integer wolf population, but as with the caribou population, we don't have to do any rounding or truncation to keep things simple. We created a Microsoft Excel spreadsheet file to calculate and display the caribou and wolf populations using these equations. This spreadsheet, caribou.xlsx, can be found on the book website (https://fractals. laplante.io). Let's use this spreadsheet to simulate the population dynamics and some of the possible chaotic behaviors.

First, by careful selection of the predator and prey populations, the K factor and the birth and death rates, the system should show stable oscillations of both species. Otherwise, the system will become unstable (chaotic), resulting in the extinction of one or the other populations.

For example, Fig. 3.1 depicts the populations over 1000 months with the following parameters:

- initial population of caribou, $caribou(0) = 10,000$
- initial population of wolves, $wolf(0) = 1,500$
- birth rate for caribou, $caribou_b = 0.01$
- death rate for wolves, $wolf_d = 0.05$
- death rate to contact ratio $K = 0.000006$

Note how this is a nicely stable system. When the caribou population gets too high, the wolf population increases shortly thereafter to keep it in check. When the caribou population drops, the wolf population falls soon after.

You will find that our little predator–prey system is not very sensitive to the initial populations (try adjusting these using the program or spreadsheet).

---

[2]As with the caribou, we ignore other means of population increase.

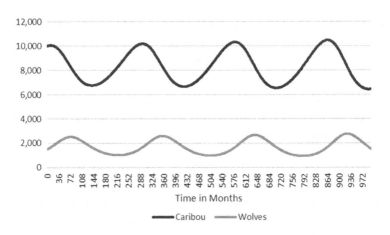

Fig. 3.1. Population dynamics of caribou–wolf system with $K = 0.000006$.

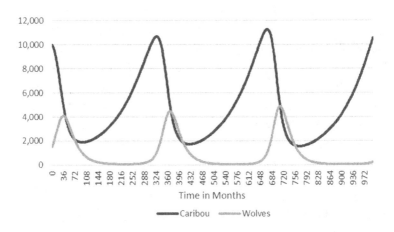

Fig. 3.2. Population dynamics of caribou–wolf system with $K = 0.000010$.

However, the system is extremely sensitive to the death rate to contact ratio, K.

For example, if we change the parameter K, the death rate to contact ratio, even slightly, to $K = 0.00001$, we generate the population profile shown in Fig. 3.2. Try tweaking the spreadsheet with these values. Note how there are wild swings in both populations, and at times, the wolf population is dangerously close to extinction.

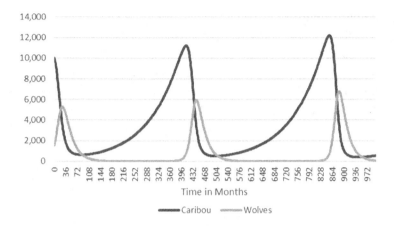

Fig. 3.3. Population dynamics of caribou–wolf system with $K = 0.000014$.

Finally, when parameter $K = 0.000014$, the system is completely unstable, as depicted in Fig. 3.3.

The initial population of caribou is quickly decimated, leading to the eventual drop in wolves. Both populations make a weak recovery, but when the caribou population drops, the wolves are eventually extinct. This leads to an explosion in caribou, which would probably drop dramatically due to lack of food, space, disease, and so on, although this is not captured by our model.

Using the spreadsheet or Rust program, you can experiment with the initial conditions and the various parameters to determine which of them lead to instability. You can also make the model more sophisticated, for example, by taking into account migration of wolves and caribou and death by other means besides wolves. These possibilities are explored in the end-of-chapter exercises.

Finally, we set up this population dynamics experiment as a *discrete simulation*, that is, we use a finite difference equation to model it.[3] We could have created a *continuous simulation*, but this would have involved a mathematical tool called a *differential equation* and very sophisticated software to solve it.

---

[3]A finite difference equation is a recursive equation that describes a function at time t in terms of the function at previous times $t - 1$, $t - 2$, and so on.

3.2   Animal Images

Fractals can be used to generate images that resemble many types of animals. For example, an infinite number of self-similar seals or dolphins can be seen frolicking in Fig. 3.4. This image was generated by running program seals.rs, with the IFS codes given in Table 3.1.

Fig. 3.4.   Seals.

Full-color image:

Table 3.1. IFS transformation rule for seals.

	1	2	3	4	5	6	Probability
1	−0.5	0	0	0.5	0	0	0.25
2	−0.5	0	0	0.5	2	0	0.25
3	−0.4	0	1	0.4	0	1	0.25
4	−0.5	0	0	0.5	2	1	0.25

```
Command line

cargo run --bin seals
```

3.3 Genetics

It is no wonder that some genetic researchers theorize that the mutation of genes, previously thought to be random, is not, but rather chaotic with various strange attractors.

Moreover, we can conjecture about the morphology of cells. Is the shape of a cell random or chaotic? Figure 3.5 shows an amoeba-like image generated from the filled Julia set of $f(z) = Z^2 + 0.3 - 4i$.

```
Command line

cargo run --bin mandel_julia amoeba
```

You can generate it yourself with the "julia_simple" program by selecting "Amoeba" for F.

3.4 Weather

It's relatively easy to find fractal-like images in clouds, storm-driven waves and hazy horizons. But, as we have observed, the weather is widely known to represent chaos. Storms and calms often appear without explanation. Embarrassed weather persons are constantly trying to decide where a certain prediction went awry. Physicist Edward Lorenz observed that, in theory,

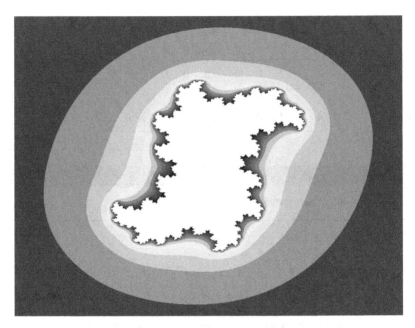

Fig. 3.5.   An amoeba-like image generated from the filled Julia set of $f(z) = Z^2 + 0.3 - 4i$.

Full-color image:

the flapping of a butterfly's wings in Tokyo might cause a storm over New York.[4]

In this sense, measuring the weather can also affect it. Certainly, the devices measuring wind speed have a more profound influence than the flapping of a butterfly's wings. One is reminded of the well-known principle in physics called the Heisenberg uncertainty. The principle states that one cannot know precisely the position and velocity of a particle at the same instant. An interpretation of this is that in measuring the position or velocity of the particle, the measuring instrumentation changes one or both. Might this not imply that by measuring the forces that determine weather, we are doomed to affect it, thus rendering our predictions hopelessly inaccurate? We further explore uncertainty in dynamical systems in Chapter 5.

---

[4]The science fiction movie *The Butterfly Effect* (2004) is about the dangerous cause–effect relationship of time travel.

3.5 Scenes from Nature

In this section, we show many beautiful computer-generated images, most of which were created by playing with the data in the transformation matrix of Barnsley's iterated function system algorithm (Barnsley, 2012).

3.5.1 *Trees, Leaves and Flowers*

Many beautiful trees, leaves and flowers can be generated using both IFS and Julia set fractals. For example, one of the most commonly seen fractals is the fern leaf[5] shown in Fig. 3.6.

Fig. 3.6. A fern leaf.

Full-color image:

---

[5]Ferns are one of the most fractal-like plants. Referring to our previous comment about fractal-like food, the furled heads of the fiddle head fern, which looks much like Fig. 3.6, are quite edible and tasty, and the Romanesco variety of cauliflower is also very fractal-like and tasty.

Fig. 3.7.   A fern leaf generated via IFS.

Full-color image:

Now, look at Fig. 3.7. You can see how it looks much like the real fern leaf.

```
Command line
cargo run --bin fern-ifs
```

This image was generated by iterating a well-known mapping rule, encoded in program fern-ifs.rs, which you can run. Table 3.2 shows a matrix encoded form of the mapping for the fern.

Table 3.2. IFS transformation rule for fern.

	1	2	3	4	5	6	Probability
1	0.5	0.0	0.0	0.16	0.0	0.0	0.01
2	0.85	0.04	−0.04	0.85	0.0	1.6	0.85
3	0.2	0.26	0.23	0.22	0.0	1,6	0.07
4	0.15	0.28	0.26	0.24	0.0	0.44	0.07

Table 3.3. IFS transformation rule for tree.

	1	2	3	4	5	6	Probability
1	0.0	0.0	0.0	0.5	0.0	0.0	0.05
2	0.42	−0.42	0.42	0.42	0.0	0.2	0.40
3	0.42	0.42	−0.42	0.42	0.0	0.2	0.40
4	0.1	0.0	0.0	0.1	0.0	0.2	0.15

Next, by changing the parameters in the IFS matrix, we can generate a tree using code similar to the one used above. In this case, the program is called tree-ifs.rs, and it uses the mapping rule described in Table 3.3.

The output of the program is shown in Fig. 3.8.

```
Command line

cargo run --bin tree-ifs
```

By outputting many trees of different size, color and position, we can create a forest. Program forest.rs does just that and its output is shown in Fig. 3.9.

Fig. 3.8.   IFS representation of a tree.

Full-color image:

```
cargo run --bin forest
```

Fig. 3.9.   A forest of randomly generated fractal trees.

Full-color image:

The program is given as follows:

```rust
use ndarray::{array, Array, Ix2};
use rand::Rng;
use rust_fractal_lab::ifs::IfsProgram;

fn main() {
    let mut program = IfsProgram::default();
    let mut rng = rand::thread_rng();

    let d: Array<f32, Ix2> = array![
        [0.0, 0.0, 0.0, 0.5, 0.0, 0.0, 0.05],
        [0.42, -0.42, 0.42, 0.42, 0.0, 0.2, 0.40],
        [0.42, 0.42, -0.42, 0.42, 0.0, 0.2, 0.40],
        [0.1, 0.0, 0.0, 0.1, 0.0, 0.2, 0.15],
    ];

    for _ in 0..150 {
        let shift_x = rng.gen_range(-0.5..0.5);
        let shift_y = rng.gen_range(-0.5..0.5);
        let scale = rng.gen_range(1.0..10.0);

        let color = {
            match rng.gen_range(0..=9) {
                // Most trees are green
                0..=7 => [0.0, 0.39, 0.0, 1.0],
                // Some trees are yellow
                8 => [0.8, 0.95, 0.0, 1.0],
                // Some trees are dead (brown)
                9 => [0.64, 0.16, 0.16, 1.0],
                _ => unreachable!(),
            }
        };

        program.sample_affine(&d, color, 2000, scale,
            shift_x, shift_y);
    }

    // Use a point size of 1.4
    program.run(Some(1.4));
}
```

Lines 17 and 18 choose a random x and y shift for each tree. Line 19 chooses a random scale factor between 100% and 1000%. Lines 21–31 choose a random color for each tree, such that about 80% of the trees are green, 10% are dead (brown), and 10% are yellow. This procedure is repeated to create 150 trees. Each tree is created by sampling the IFS 2,000 times.

Another type of forest is illustrated in Fig. 3.10 and can be regenerated by running the program redmoscl.rs.

```
Command line
cargo run --bin redmoscl
```

Here, we see a view of a redwood forest, with a lush green floor and huge trees, whose tops are obscured by a mist. The effect was also achieved using the iterated function system algorithm. The floor of the forest is simply composed of our trees again, while the redwoods and mist were generated with other IFS with different parameters.

Finally, green seaweed was produced using the program seaweed.rs. The output is shown in Fig. 3.11.

```
Command line
cargo run --bin seaweed
```

The IFS codes for the seaweed are given in Table 3.4.

Using Julia sets, beautiful flowers can be created. For example, a Julia set of

$$f(z) = z^2 + 0.384$$

gives the lovely four-petaled rose, as shown in Fig. 3.12.

```
Command line
cargo run --bin mandel_julia flower1 -c inferno
```

Fig. 3.10.   A redwood forest.

Full-color image:

A chrysanthemum flower is generated by changing the constant term slightly to 0.2541, that is,

$$f(z) = z^2 + 0.2541$$

Fig. 3.11. Green seaweed.

Full-color image:

The output is shown in Fig. 3.13.

```
Command line

cargo run --bin mandel_julia flower2 -c inferno
```

Table 3.4.   IFS transformation rule for seaweed.

	1	2	3	4	5	6	Probability
1	0.5	0	0	0.5	0	0	0.25
2	0.5	0	0	0.5	2	0	0.25
3	0.4	0	1	0.4	0	1	0.25
4	0.5	0	0	0.5	2	1	0.25

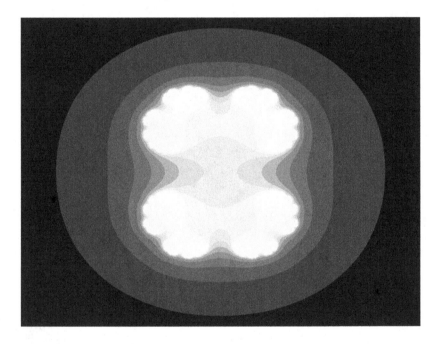

Fig. 3.12.   Four-petaled flower from the Julia set of $f(z) = z^2 + 0.384$.

Full-color image:

3.5.2 *Clouds*

The billowy appearance of some fractals can be exploited to generate cloud-like pictures. For example, Fig. 3.14 shows a threatening storm cloud.

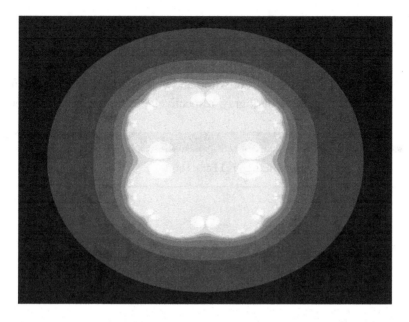

Fig. 3.13. Another four-petaled flower from the Julia set of $f(z) = z^2 + 0.2541$.

Full-color image:

Fig. 3.14. A fractal storm cloud.

Full-color image:

The program finds the Julia set of

$$f(z) = z^2 - 0.194 + 0.6557i$$

and suppresses all colors except white and yellow which are mapped into the colors light gray and dark gray, respectively, using the case statement.

Command line

```
cargo run --bin mandel_julia cloud
```

```
1   subroutine(colorize_t)
2   vec4 ColorizeCloud(uint i) {
3       switch (i / 2u) {
4           // light grey
5           case 4u: return vec4(0.83, 0.83, 0.83, 1);
6           // dark grey
7           case 5u: return vec4(0.39, 0.39, 0.39, 1);
8           // white
9           default: return vec4(1, 1, 1, 1);
10      }
11  }
```

We can also use IFS systems to generate three-dimensional appearing clouds. For example, the clouds shown in Fig. 3.15 were generated with the program clouds.rs. The IFS codes for it are given in Table 3.5.

Command line

```
cargo run --bin clouds
```

3.5.3 *Rocks and Boulders*

The generation of rocks and boulders can be handled similarly by using the cloud generation code and changing colors. For example, try changing colors of the FCloud subroutine (for mandel_julia.rs) by mapping white into brown and yellow into red to generate a rock-like formation.

Fig. 3.15. "Three-dimensional" fractal clouds.

Full-color image:

Table 3.5. IFS codes for three-dimensional fractal clouds.

	1	2	3	4	5	6	Probability
1	0.5	0	0	0.5	0	0	0.25
2	0.5	0	0	0.5	2	0	0.25
3	−0.4	0	1	0.4	0	1	0.25
4	−0.5	0	0	0.5	2	1	0.25

Fig. 3.16.   Fractal rocks using same IFS codes as the clouds.

Full-color image:

Another way to generate a rock formation is with iterated function systems. For example, consider the IFS code table for the program clouds.rs. By changing the color to brown, the rocks shown in Fig. 3.16 are generated.

3.5.4 *Snowflakes*

We can generate images that look like snowflakes easily using fractal techniques. One type of snowflake, shown in Fig. 3.17, was generated from the Julia set of

$$f(z) = z^2 + 0.11031 - 0.67037i$$

Command line

```
cargo run --bin mandel_julia snowflakes
```

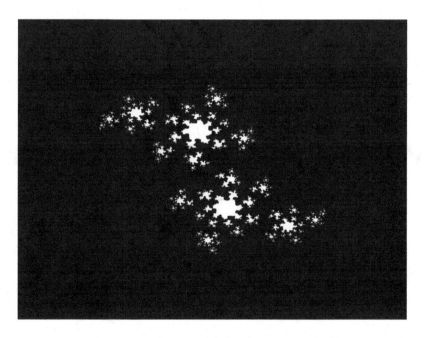

Fig. 3.17.   Snowflakes generated by mandel_julia.rs.

Full-color image:

This marvelous effect is achieved by coloring only the white pixels, with the following code:

```
1   subroutine(colorize_t)
2   vec4 ColorizeSnowflakes(uint i) {
3       if (i >= 8u) {
4           return vec4(0, 0, 0, 1);
5       } else if (i >= 12u) {
6           return vec4(1, 1, 1, 1);
7       }
8
9       return vec4(0, 0, 0, 1);
10  }
```

Finally, a lovely snow fall, as shown in Fig. 3.18, was generated by repeated random generation of the "cross fractal," much in the same way the forest was generated. Look at one of the snow flakes. It is just a square divided into nine equal parts with the four outer-middle boxes removed, as shown in Fig. 3.19.

```
Command line

cargo run --bin fall
```

The cross fractal is generated with the IFS codes shown in Table 3.6. Many of these little fractals are generated in different scales and positions to achieve the effect. You can produce the snow fall by running the program fall.rs. The resultant image appears to progress from a flurry to a blizzard.

3.5.5 *Galaxies*

Slight modification of the program fall.rs yields what appears to be the view of some unknown region of space shown in Fig. 3.20. By changing

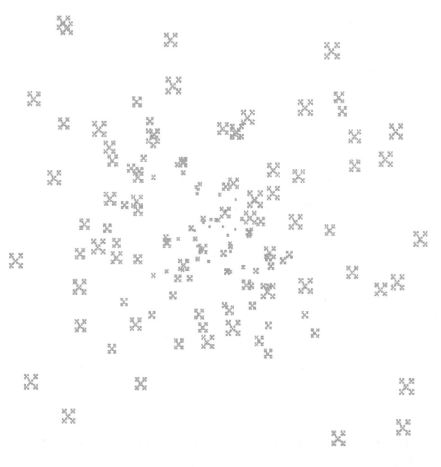

Fig. 3.18. A snow fall generated by fall.rs.

Full-color image:

the snowflake scaling factor so that it is very small, the flakes become stars. This can be seen by running the program galaxy1.rs.

```
Command line
cargo run --bin galaxy1
```

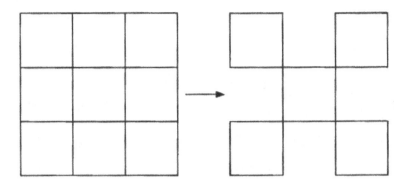

Fig. 3.19.   Cross fractal used to generate snowflakes.

Table 3.6.   IFS codes for cross fractal.

	1	2	3	4	5	6	Probability
1	0.33	0	0	0.33	1	1	0.20
2	0.33	0	0	0.33	10	1	0.20
3	0.33	0	0	0.33	1	10	0.20
4	0.33	0	0	0.33	10	10	0.20
5	0.33	0	0	0.33	5	5	0.20

Finally, looking again at Fig. 3.17, we see that it resembles twin swirling galaxies.

3.5.6   *Coastlines*

Any one of the fractals generated using Julia sets, for example, Figs. 3.14 and 3.22, could represent the coastline of some mythical country viewed from above. Since the fractal is self-similar, it doesn't matter whether the viewer is 1,000 miles or 1 foot away.

In addition, any section of these fractals has the property that it has an infinite length. This may defy intuition, but if you tried to use a piece of string to trace the coastal outline, you would run out of string. Perhaps, the coast of England is infinitely long, as Mandelbrot has said.

Fig. 3.20. A randomly generated view of space.

Full-color image:

3.6 Fractals in the Human Body

Many structures within the human body suggest a complex interrelation between biological development, form and function. Scientists have wondered if underlying physical constraints lead, through scaling, to the ultimate form of plants and animals. For example, does the shape of a DNA molecule have direct relationship to the shape of the organism it describes? Let us look at some instances where the human body may harbor fractals.

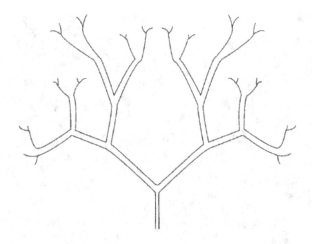

Fig. 3.21. The bronchial tree structure in our lungs resembles a fractal.

3.6.1 *Bronchial Growth*

Our lungs contain millions of air sacs called *alveoli*, which provide a mechanism for the exchange of gases. These are connected via increasing larger *bronchial tubes* to the trachea in a structure shown in Fig. 3.21, which is very similar to the tree fractal shown in Fig. 3.8.

As the bronchial tree branches out, its tubes decrease in size. From one branching to the next, the diameter decreases at about the same ratio until there is a change in the mechanism of flow from minimum resistance near the beginning to molecular diffusion within the alveoli. This structure may also be similar to neural connections in the brain.

3.6.2 *Neuron Growth*

Some researchers have suggested modeling the wiring of the brain and neuron growth using fractal bifurcation patterns (DeAngelis, 1993). For example, the tree fractal has been cited as one mechanism for neuron wiring.[6] In addition, neural activity tends to be fractal-like and chaotic — more so when the brain is involved in active problem solving (DeAngelis, 1993).

---

[6]The same mechanism has been investigated as a model for wiring local phone systems.

Interestingly, with fractals, we can generate a structure that resembles the main connectors between the neural processing elements of the brain called *dendrites*. For example, look at the filled Julia set generated by the function

$$f(z) = z^2 + i$$

depicted in Fig. 3.22.

Command line

```
cargo run --bin mandel_julia dendrite
```

Could the function f be the underlying mathematics behind the dendrite? Nobody knows.

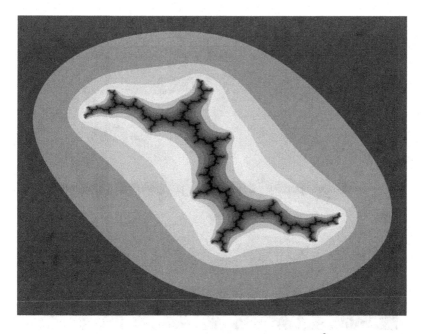

Fig. 3.22. Dendrite structure generated by $f(z) = z^2 + i$.

Full-color image:

3.6.3 *Physiological Processes*

In addition to physical structure, physiological processes may be subject
to the scaling properties that characterize fractals. Fractal processes within
organisms cannot be characterized by a single scale of time, but instead
have components at many frequencies. For example, some researchers have
related the geometry of the nerves in the heart to the associated electrocardiogram output. Similar findings have been reported for the electrical
activity of a neuron and variability in heart rate. Some scientists even spec-
ulate that diseases are caused by a disruption of the normal fractal scaling
(West and Goldberger, 1987).

Recall we were able to generate an EKG-like image using wavelets in
Chapter 2. The image in Fig. 3.23 generated by the program mandel_julia.rs
using the Julia set for

$$f(x) = z^2 - 1.5$$

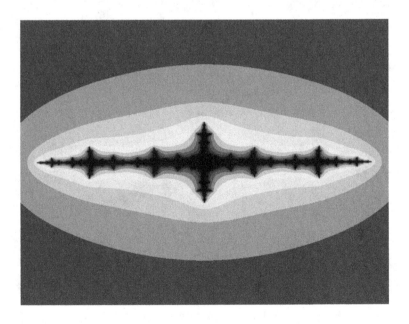

Fig. 3.23.   An EKG output simulated using a Julia set.

Full-color image:

also looks somewhat like EKG output, suggesting that fractals could also be used for storing and compressing real EKGs.

```
Command line

cargo run --bin mandel_julia ekg
```

3.6.4 *Chaos of the Mind?*

Theories in psychology (DeAngelis, 1993; Moran, 2018) conjecture that behavior may be determined by chaotic phenomenon (some persons believe this already). For example, viewing the mind as a complex dynamical system, psychologists contend that everyday, "normal" behavior represents attracting states. However, the chaotic and unstable nature of the mind often leads to drastic, random behavior shifts, which when harmless are considered impulsiveness, but when harmful are considered dangerous psychosis.

Some theorists believe that "crisis-prone" families are not behaving randomly but simply according to a different norm. Practitioners believe that because the behavior patterns are highly chaotic, they can be changed from an abnormal one to a normal one with only a slight nudge.

3.7 Exercises and Things to Do

Exercise 3.1

Describe another natural phenomenon (not mentioned in this chapter) that could be fractal-like in nature.

Exercise 3.2

Find or take an image for any natural phenomenon not described in the book. Now, using the IFS code, experiment with different parameters to try to replicate the image (we know it's hard using trial and error).

Exercise 3.3

Take a walk through a garden or woods or field and find something that is fractal-like and take a picture of it. Now, using any of the fractal generation techniques that we have studied, create a fractal that looks similar to the picture.

Exercise 3.4

Create a number of fractals using any technique discussed and then, using an image editing tool, combine the fractals to build a "garden."

Exercise 3.5

For each of the following pairs, identify predator–prey combinations and settings in which the population dynamics might behave similar to the wolf–caribou pair:

- (insect,insect)
- (insect,plant)
- (mammal,fish)
- (fish,fish)
- (bird,fish)
- (bird,mammal)

For example, for (bird,fish) in a particular lake, you might have a similar population dynamic for osprey and bass.

Exercise 3.6

Implement the caribou–wolf population simulator in Rust (or the language of your choosing). You will need inputs to represent initial population counts, caribou birth rate, wolf death rate, contact–death ratio, and number of months to simulate. A basic version of this program could just hardcode those values, while a more sophisticated one could either prompt the user to enter them or take them as command-line arguments.

There is no need to use OpenGL or shaders. If you want to draw a plot, consider using the imgui-rs crate.

Exercise 3.7

Try running either your Rust caribou–wolf population simulator or the Excel spreadsheet using different values of caribou birth, wolf death, and encounter (K) parameters and different initial populations to get different results. Write a one-page paper, summarizing your findings.

Exercise 3.8

Modify the Rust program or Excel spreadsheet for the caribou–wolf population dynamics to account for various other factors, such as a different wolf birth rate (not based on caribou deaths), migration of wolfs and/or caribou, and deaths from other factors (such as aging, disease and accident). This program can become very complicated, but the more factors you include, the more realistic the model.

Exercise 3.9

See if you can find some Romanesco cauliflower or fiddle head ferns at a specialty market. While these are not available everywhere, nor year round, they are relatively common in some areas and some stores. Find a recipe for either (any cauliflower recipe will work with Romanesco), cook and enjoy the result.

Chapter 4

Chaos and Fractals in Human-Made Phenomena

"Chaos umpire sits,
And by decision more embroils the fray
By which he reigns: next him, high arbiter,
Chance governs all."

— John Milton, *Paradise Lost*

In this chapter, we focus on how human-made systems can exhibit chaotic behavior. We see how this chaotic behavior can be harnessed and how it can be harmful. We also cover how fractal images can be used in modeling and predicting the performance of human-made systems.

4.1 Turbulent Flow

Anyone who has ever flown in an airplane is familiar with turbulent flow or turbulence. The fluid (air and liquids) flow is a major application area in dynamical systems. *Turbulence* is characterized by disorder on all scales, with backward eddy currents and circular waves. In most systems, it is undesirable, creating drag and loss of energy through increased friction. Many human-made systems can exhibit chaotic turbulent flow: from the output of jet engines to the flow of oil through a pipeline. Automobile, airplane, and boat manufacturers use wind tunnels to design vehicle profiles that do not promote turbulence.[1]

From everyday life, one can readily find situations that exhibit turbulent behavior. For example, boil a pot of water over a stove. As the water slowly boils, steam begins to escape from the surface. Next, the water slowly and rhythmically begins to ripple until finally a turbulent, rolling boil is reached. This turbulence is chaotic and random appearing, yet there is some semblance of regularity as well. It's almost as if some pattern wants to emerge.

A second demonstration of turbulence, suggested by Moon (1982) also involves water. Take a dinner plate and place it under a tap. Fill the dish with water to overflowing level and continue to gently run the water. Place a ping pong ball in the dish, and adjust the water until the ball bounces around merrily, performing chaotic oscillations.

4.2 Structures

Escher was famous for his sketches and woodcuts that depicted impossibly beautiful buildings and other architectural marvels. The architect, Pei, was also noted for his impossible designs, with bizarre acute angles that defined the conventional technique. Find a book on building architecture to see the mathematical intuition which Pei possessed. In the spirit of Escher and Pei, using fractals, and in particular IFS fractals, we can create artificial structures that appear to be functional as well as beautiful.

For example, try running the program castle.rs, which is shown in Fig. 4.1. It appears to be the walled ramparts of some medieval castle. The IFS codes for the program are shown in Table 4.1.

```
Command line

cargo run --bin castle
```

---

[1]Turbulent flow is found in many natural settings, for example, waterfalls and crashing waves are clearly turbulent, but we discuss this in the context of human-made situations here.

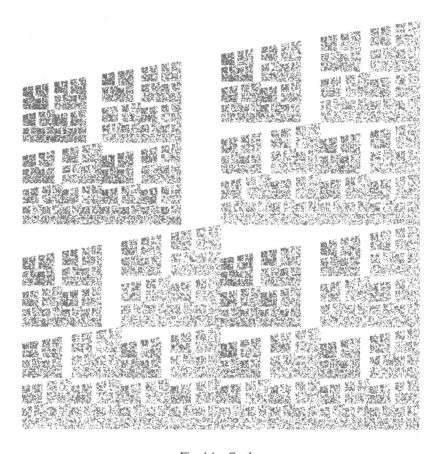

Fig. 4.1. Castle.

Full-color image:

Table 4.1. IFS codes for castle.

	1	2	3	4	5	6	Probability
1	0.5	0	0	0.5	0	0	0.25
2	0.5	0	0	0.5	2	0	0.25
3	0.4	0	0	0.4	0	1	0.25
4	0.5	0	0	0.5	2	1	0.25

The beautiful maze-like structure, shown in Fig. 4.2, can be generated by running program maze.rs. The IFS codes for this program are shown in Table 4.2.

```
cargo run --bin maze
```

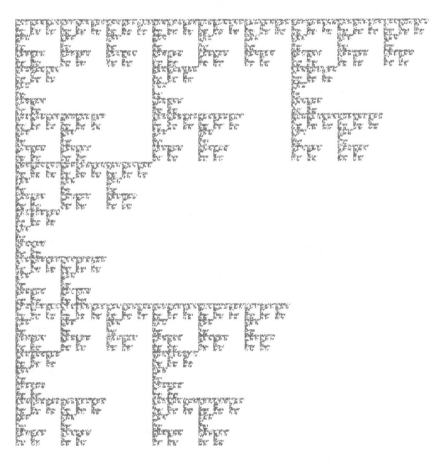

Fig. 4.2. A fractal maze.

Full-color image:

Table 4.2. IFS codes for maze.

	1	2	3	4	5	6	Probability
1	0.33	0	0	0.33	1	1	0.166
2	0.33	0	0	0.33	100	1	0.166
3	0.33	0	0	0.33	1	100	0.166
4	0.33	0	0	0.33	100	200	0.166
5	0.33	0	0	0.33	200	200	0.166
6	0.33	0	0	0.33	1	200	0.166

4.3 Computer Scene Analysis

Scene analysis is the process of extracting specific features from a larger picture or scene. Many applications exist for scene analysis, for example, mobile robots typically use scene analysis to navigate over terrain, target acquisition programs used in civilian and defense systems use scene analysis to locate a designated object inside a scene, and medical diagnosis software use feature analysis to locate specific cell configurations, such as tumors or fractures.

Fractal models are useful tools in certain types of three-dimensional scene analysis of two-dimensional images because they can model an object's surface roughness. Fractals are especially well suited for this because surface roughness is known to be scale-invariant within the effective resolution of most imaging devices.

Fractal models have been used to classify natural textures, such as skin, rock, cloth, and grass, very accurately.

4.4 Image Compression

Image compression is the process of reducing the amount of stored information needed to reproduce an image. In fractal compression techniques, the bit-by-bit storage of the image is replaced by its representation by an iterated function which requires significantly less storage. The disadvantage of course is that it often requires significant time to regenerate the image from the iterated function rather than simply displaying the image pixel by pixel.

The quality or level of compression is expressed in terms of a compression ratio. The *compression ratio* is the ratio of the bytes required to store an uncompressed image to those needed to store the compressed equivalent. Compression rates for fractal compression seem to be in the range from 20:1 to 60:1 but with questionable quality.

To illustrate the power of fractal compression, consider the forest that we generated in Fig. 3.9. If the computer screen that displays it is 640 by 480 pixels and requires 16 bits or 2 bytes per pixel, then

$$640 \times 480 \times 2 = 614400$$

bytes of storage are needed, whereas the program used to generate it required only the data contained in the IFS matrix. Assuming that each number in the matrix required 4 bytes and four additional 4-byte numbers were needed for the probability that the transformation in a row was applied, then we only needed to store

$$24 \times 4 + 4 \times 4 = 112$$

bytes. Then the compression ratio for the forest image is

$$\frac{\text{Number of bytes for screen image}}{\text{Number of bytes for IFS codes}}$$

or $\frac{112}{614400}$, which is a compression ratio of 5485:1. This is an amazing savings in storage, which is due to the very low quality of the image rendered. A higher quality image would necessitate a much lower compression ratio. Finally, if the image were to be transmitted by a satellite to the earth, an incredible savings in the time needed to transmit the image is also realized.

According to Barnsley (2012), fractal compression is facilitated by a measure of deviation between a given image and its approximation by an iterated function system (IFS). The Collage Theorem (Barnsley, 2012) essentially states that to find an IFS attractor that is close to the desired image, one must find a set of mappings and transformations such that their union or collage is close to the desired image. This process of finding the transformations is unfortunately agonizingly slow even on the most powerful supercomputers.

4.4.1 *Problems with Fractal Compression*

One of the greatest problems with fractal compression of images is that it is difficult to find single fractals that will closely represent an arbitrary image. But there are certain fractals that are known to reproduce natural images, such as trees, mountains, clouds, and so on. Even for more scenes that are relatively uniform, it is possible to find a single fractal that can be tuned to compress the scene.

For example, look at Fig. 4.3, a picture of a pond.

Fig. 4.3. "Winter reflections" by waferboard is marked with CC BY 2.0. To view the terms, visit https://creativecommons.org/licenses/by/2.0/?ref=openverse. Originally in color.

Full-color image:

Parts of this image certainly appears to be fractal-like. In fact, by tinkering with various IFS parameters, we obtain the suggestive image shown in Fig. 4.4.

```
Command line

cargo run --bin swamp
```

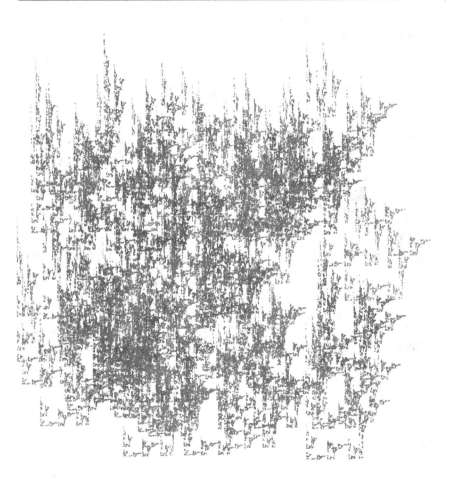

Fig. 4.4. Computer-generated equivalent of swampy pond shown in "winter reflections."

Full-color image:

Table 4.3.   IFS codes for one clump in a swamp.

	1	2	3	4	5	6	Probability
1	0.5	0	0	0.25	1	1	0.25
2	0.25	0	0	0.7	50	1	0.25
3	0.25	0	0	0.7	1	50	0.25
4	0.5	0	0	0.25	50	50	0.25

It certainly looks similar to the upper half of Fig. 4.3.

To generate this image, we use the IFS codes given in Table 4.3 and generate this fractal in many positions, sizes and colors similar to the forest image.

The code for this fractal is contained in swamp.rs. You can play with these parameters further to try and more closely match the photo.

But for arbitrary images that include more than just trees and clouds, how can fractals be used to compress them? The answer is that the image is divided into a number of subimage blocks (usually ranging from 2 × 2 to 16 × 16). Then through an iterative process, an attractor is found for each block such that the error (difference between the original and compressed block images) is minimized to some arbitrary amount.

A significant problem associated with the compression of images using fractals or any method is that there is a time penalty assessed during both the compression and decompression stages. This penalty can sometimes be particularly onerous for fractal compression of complex images that require numerous blocks. In cases involving real-time image processing, where in order for the human eye to perceive continuous motion, the screen must be updated at approximately 33.3 times per second, this decompression delay may be unacceptable.

4.5   Economic Systems

Mandelbrot was one of the first to recognize that scaling is an important feature of pricing in economics. He analyzed the price of cotton (based on the Department of Agriculture figures) over the period of time from 1880 to 1958. An interesting pattern emerged. When Mandelbrot plotted a suitable

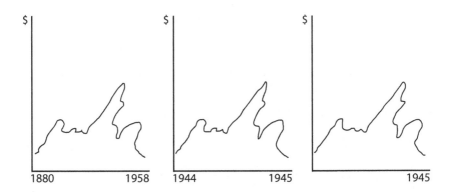

Fig. 4.5. Self-similarity of cotton prices over centuries, years, and months.

function of the price of cotton over the period 1900–1905, it resembled the same plot for the periods 1880–1940 and 1944–1958. In other words, he identified a self-similarity in price across the timescale (see Fig. 4.5).

However, it is still unknown if Mandelbrot discovered fundamental truth in pricing of commodities or if it was simply a coincidence. Obviously, if one could guarantee the fractal nature of the price of cotton with certainty, then they should be making a killing on the Chicago Mercantile Exchange!

However, to see that this may be possible, consider a simple economic system that features a single product, say widgets, with price P and a market of buyers and sellers. If the amount of widgets produced stayed fixed, then basic economic theory says the price should be a linear function of the demand and thus would rise or fall by a factor of a. The price of widgets at time t would then be

$$P(t + 1) = aP(t)$$

That is, the price at any time $t + 1$ is just the price at the previous time times the factor a.

Suppose that, in order not to appear greedy, the sellers decide to lower their price by the quantity $aP(t)^2$ (the sellers know that what they lose in profit margin, they make up in volume anyway). Then the price at time $t + 1$ is given by

$$P(t + 1) = aP(t) - aP(t)^2 \tag{4.1}$$

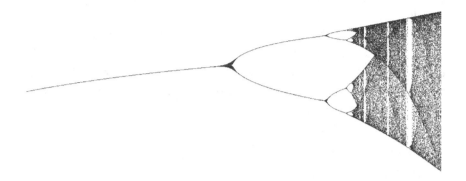

Fig. 4.6. Bifurcation diagram for model economic system.

Full-color image:

This equation is known as the *logistics equation*, and it was first proposed as a model for population growth by Verhulst[2] in 1845.

Let's simulate the system by assuming that the starting price for the widget is 90 cents and sweeping the demand factor a from 2.5 to 4.0. We then iterate the price over time (the timescale could be days, months, years, or otherwise), skipping the first 50 iterations to allow the price to "stabilize." We then plot the price $P(t)$ on the y-axis against the demand factor a on the x-axis. You can see this by running program bifurcation.rs in a special "price model" mode. Its output is shown in Fig. 4.6.

```
Command line

cargo run --bin bifurcation -- --price
```

Note that the fractal is similar to the bifurcation fractal shown in Fig. 1.4. Also note that for any particular value of a (plotted on the x-axis), there are many prices associated with it, that is, the price is unstable. However, there

---

[2]Pierre François Verhulst (1804–1849) was a Belgian mathematician best known for his work in modeling population growth.

are a few values for which the price seems to flip-flop between two numbers. These bands of stability are characterized by bald spots in the plot.

One final note is that equation (4.1) is almost identical to equation (2.11) used to generate the Mandelbrot set except that the former involves real numbers instead of complex ones. It should be no wonder then that they both generate self-similar images.

4.6 Cellular Automata

A type of mathematical abstraction, called *cellular automata*, have a profound relationship to both fractals and chaos. Cellular automata have been studied as a model for biological cell behavior and massively parallel computers.

Cellular automata, which were originally investigated by von Neumann,[3] consist of a space of unit cells. These cells are initialized with some value, generally a "1" representing a "live" cell and a "0" for a dead or unoccupied cell. Different characters can be displayed to represent these states, but the idea is that some rule describing the evolution of the system is defined. This rule describes the contents of a unit cell at time t in terms of the contents of the cell and its neighbors at time $t - 1$.

An important feature of cellular automata is the ability to self-organize or, in the terms of chaos theory, find attractors. In addition, many types of cellular automata will eventually attract to stable, fractal-like formations. This attraction occurs, with relative indifference to the initial state of the cell field.

---

[3]John von Neumann (1903–1957) was a Hungarian-American physicist and mathematician who made significant contributions to dynamical systems, economics, mathematics, computer science and many other areas. His deep contributions to so many fields rank him among the greatest scientists of all time. Due to his work in defining the "standard" architecture for early computers, the most common form is termed as "von Neumann" architecture, but there are many other theorems and principles named after him.

Stephen Wolfram[4] classified cellular automata in a way that helps to reveal their relationship to chaos (Wolfram, 2002).

Class I: evolution to a homogeneous state (an attractor),
Class II: evolution to isolated periodic segments,
Class III: evolution which is always chaotic,
Class IV: evolution to isolated chaotic segments.

We will soon see some cellular automata that fit each of these categories. As we go along, try to decide for yourself.

4.6.1 *One-Dimensional Cellular Automata*

A cellular automaton where the cells are organized in rows and a cell's contents at time t are based only on the contents of the cell and its neighbors on either side at time $t - 1$ is called a *one-dimensional cellular automaton*. In one-dimensional cellular automata, we trace the evolution of the system by observing the row at time t followed by the row at time $t + 1$ and so on. In many cases, the result is chaotic or unstable, but in some cases, a strange attractor is found.

For example, the program 1d_life.rs implements a cellular automaton that follows this cell rule:

$$a_0' = (a_{-1}a_0a_1) + (\overline{a_{-1}}a_1) + (a_0a_1) \tag{4.2}$$

Let's see what this rule means.

The symbol a_0' represents the contents of a given cell at time t. Similarly, a_0 is the contents of the cell at the previous time, $t - 1$. Finally, a_{-1} is the contents of the cell on the left at time $t - 1$ and a_1 is the contents of the cell on the right at time $t - 1$.

---

[4]Stephen Wolfram, born in 1958, is a British-American mathematician and physicist who has made important contributions to cellular automata and symbolic computation. He is the founder of the company Wolfram Research, which created the important computational engine, Mathematica, and the symbolic computation system, Alpha, which can be used to solve or check many of the computational exercises in this book. You can find Alpha at https://www.wolframalpha.com.

The multiplication symbol, $\cdot$, represents the *boolean AND operation*, which produces a one only if both operands are one.[5] The addition symbol, $+$, represents the *boolean OR operation*, which produces a one if one or both operands are one. Finally, the bar over a cell contents, for example, $\overline{a_0}$, indicates that its *Boolean complement* is to be taken, which is simply a one if the cell contains a zero and vice versa. Thus, the cell rule says the following in words:

> A cell is alive if both its neighbors and it are alive, or if its left neighbor is dead and its right neighbor alive, or if it and its right neighbor are alive.

Isn't the mathematical notation more compact?

To run 1d_life.rs, give it an initial line of asterisks ("*") corresponding to the live cells. Try running the command given in command line 4.6.1.[6]

Command line 4.6.1: Sierpinski triangle

```
cargo run --bin 1d_life "40[ ]*40[ ]"
```

The output of the program should look like a Sierpinski triangle, as shown in Fig. 4.7.

It is amazing that like a seed crystal, a single cell site results in the strange attractor of the Sierpinski triangle.

You can have fun experimenting with 1d_life.rs. The exercises at the end of the chapter provide you with some ways to try this. To change the rule that is applied, you need to implement the `Rule` trait. The trait is given as follows:

---

[5]Boolean operations are intended to be applied to binary variables and constants. *Binary variables and constants* can only take on the values 0 or 1.

[6]The part in quotes is shorthand that means: 40 spaces, followed by an asterisk, followed by 40 more spaces.

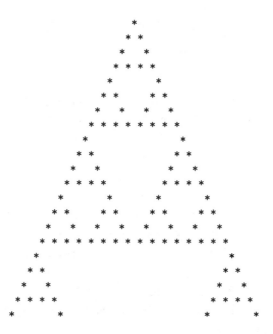

Fig. 4.7. Sierpinski triangle output of program 1d_life.rs.

```
/// Represents a rule to be applied to the state
trait Rule {
fn apply(&self, a_neg1: bool, a_0: bool, a_1: bool) ->
↪  bool;
}
```

The default cell rule (corresponding to Rule 4.2 above) is as follows:

```
struct Rule1;
impl Rule for Rule1 {
    fn apply(&self, a_neg1: bool, a_0: bool, a_1: bool) ->
↪  bool {
        (a_neg1 && !a_0 && !a_1) || (!a_neg1 && a_1) || (a_0
↪  && a_1)
    }
}
```

Once you have implemented your rule, just change the line in `main` that chooses the default cell rule. For example, if we implement some rule called `MyRule`, then we modify line 3 as follows:

```
let rule: Box<dyn Rule> = match args.rule {
    // By default, the below rule is selected
    None => Box::new(MyRule::default()),
    Some(rule) => {
        match rule {
            BuiltinRule::Rule1 => Box::new(Rule1::default()),
            BuiltinRule::Rule2 => Box::new(Rule2::default()),
        }
    }
};
```

Some one-dimensional cellular automata, for instance, the one we just saw, generate interesting fractal-like patterns based on an initial configuration using only one cell. These systems may be models of crystal growth in certain types of structures that are generated from a single seed crystal. However, other cellular automata take a random, multiple cell input and organize them to find attractors after many iterations. The organization may be in the form of a regular or fractal-like pattern, or it may result in some form of oscillating structure.

Let's look at one rule that organizes a random initial cell configuration into a fractal-like state. 1d_life.rs includes Rule2:

```
struct Rule2;
impl Rule for Rule2 {
    fn apply(&self, a_neg1: bool, a_0: bool, a_1: bool) ->
    ↪  bool {
        (a_neg1 && !a_0 && !a_1) || (!a_neg1 && a_1)
    }
}
```

It is essentially the same as Rule1 (and Rule 4.2), except with the evolution rule

$$a_0' = (a_{-1}\overline{a_0 a_1} + \overline{a_{-1}}a_1)$$

Run it, giving it any random (non-empty) initial cell configuration, such as

```
*** *    **   * **** * * *** ***   * * *** *  * ***  ** **
*** **    ** **
```

Be sure to populate the input line with many live cells — remember this is supposed to be a random starting configuration, so that asterisks or live

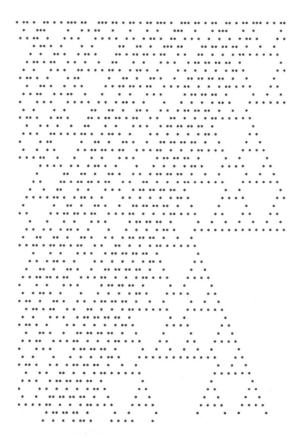

Fig. 4.8. Sample output from 1d_life.rs using `Rule2`.

cells are just as likely to occur as dead or blank cells, the same configuration will eventually be achieved.

As a shorthand, you can use `80[?]` to represent 80 randomly initialized cells. Try the command given in command line 4.6.2. It uses the `-i` flag to run for 50 iterations.

```
Command line 4.6.2: Randomized start cells

cargo run --bin 1d_life "80[?]" -r rule2 -i 50
```

The output will resemble Fig. 4.8, but will be different each time you run it. Note that although it is random, it appears to be organizing or at least oscillating through different configurations. Try running this automaton for several hundred iterations.

4.6.2 *Two-Dimensional Cellular Automata*

A cellular automaton that is organized as a two-dimensional matrix or array of cells is called a *two-dimensional cellular automaton.* Here a cell's contents at time t is based on its own and the contents of all its immediate neighbors at time $t - 1$. Cellular automata have many applications in cryptography, image and texture analysis, and in modeling of infectious disease transmission and urban and environmental growth (Hadeler, 2017).

A particular two-dimensional cellular automaton that has been studied extensively is the "Game of Life," developed by Conway.[7] In the Game of Life, the local rule states that a cell "dies" (gets a value of zero) unless two or three of its neighbors are alive (have a value of one). If two neighbors are alive, the value of the cell site is unchanged. If three neighbors are alive, the site always takes on the value one.

Depending on the initial configuration, various static equilibrium states, such as squares or hexagons, can arise. Oscillating or periodic segments can exist as can "travelling" or "glider gun" states where cell configuration move across the cell field and can be regenerated indefinitely.

Program game_of_life.rs is an implementation of the Game of Life. We have made the traditional assumption that if the number of live cells around a given cell is greater than 3, then the cell dies of overcrowding. An interesting aspect of the program is that the Game of Life rules run entirely on the GPU using fragment shaders. The code is a bit too complex to cover here, but you are encouraged to take a look at the extensively commented source code if you are interested.

Try running the program. It initializes with a random starting population. If you'd like, you can even click anywhere on the screen to manually set

---

[7]John Conway (1937–2020) was a British mathematician and professor at Princeton University. He made important contributions to number theory, game theory, geometry, and many other areas.

Fig. 4.9. Game of Life.

Animation:

cells as "alive." Figure 4.9 depicts an instance of the program shortly after it was launched.

```
Command line
cargo run --bin game_of_life
```

Look for stable configurations such as squares and circles, and bi-stable configurations called "blinkers," which alternate between two states. Other configurations of cells will appear to walk across the screen. Still others will split or join. Figure 4.10 shows an instance that ran for many iterations

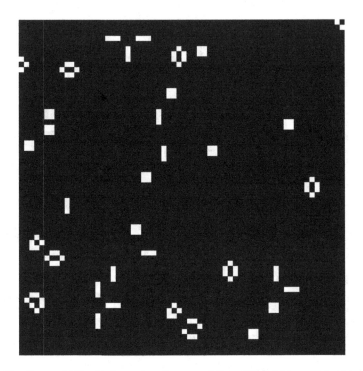

Fig. 4.10.   Game of Life after many iterations consisting of just stable and bi-stable formations of cells.

before settling into a steady state with both stable and bi-stable formations.[8] It really is a fascinating program to watch.

There are also many implementations of the Game of Life on the Web. For example, https://playgameoflife.com and at Wolfram Research's https://demonstrations.wolfram.com/RealTimeSimulationOfTheGameOf Life.

Another great resource is https://conwaylife.com/wiki/Main_Page. There you will find thousands of articles on all aspects of the Game of Life.

---

[8]Obviously, printed text can't show motion, so you'll just have to take our word for it.

4.7 Exercises and Things to Do

Exercise 4.1

Describe another kind of turbulent behavior that you have observed besides those described in the book.

Exercise 4.2

Try using Wolfram Alpha (https://www.wolframalpha.com) to solve (or check your answer) for some of the computational exercises in Chapters 1–3.

Exercise 4.3

Rewrite the one-dimensional celluar automata program so that single pixel activations represent cells.

Exercise 4.4

Implement a rule using the `Rule` trait in the one-dimensional celluar automata program to try the following rules:

(1) $a_0' = (\overline{a_{-1}\,a_0}) + (a_{-1}\overline{a_1}) + (\overline{a_{-1}}a_1)$
(2) $a_0' = (a_{-1}\overline{a_0\,a_1}) + (\overline{a_{-1}}a_0) + (\overline{a_{-1}}a_1) + (a_0 a_1)$
(3) $a_0' = (a_{-1}\overline{a_0\,a_1}) + (\overline{a_{-1}\,a_0}a_1)$
(4) $a_0' = (\overline{a_{-1}\,a_0}a_1)$

For each, run the program and use the same input as before, that is, seed it with a single cell site.

Exercise 4.5

Using any implementation of the Game of Life, produce an initial state that generates the following:

- glider gun
- a blinker

Exercise 4.6

Write a one-page biography of John von Neumann or Stephen Wolfram.

Exercise 4.7

Write a short biography of Escher or Pei with particular focus on the fractal nature of their work. Include images of their work.

Chapter 5

Dynamical Systems and Systems Theory

"We ourselves are so-called non-linear dynamical systems… I don't feel quite so pathetic when I interrupt a project to check on some obscure web site or newsgroup or derive an iota of cheer by getting rid of a pocketful of change."

— John Allen Paulos

In Chapter 1, we introduced the notion of dynamical systems as systems that have special time dependencies. That is, a dynamical system is one that describes the time dependence of a point in a geometrical space. We related these points to various chaotic phenomena and, in some cases, their fractal representations. Of course, dynamical systems and chaotic systems are closely related. For example, recall the bifurcation diagram representing the population dynamics of wolves and caribou over time discussed in Chapter 3. But dynamical systems govern the behavior of so much more, including many of the things we encounter in life.

So, in this chapter, we take a somewhat deeper dive into dynamical systems, general systems theory and related concepts with a focus on how they appear and influence our everyday lives. A comprehensive study of these topics is not our intention — we only introduce enough concepts to help further inform previous discussions to get the reader thinking and motivate further study.

5.1 Basic System Theory

A *system* is a mapping function of a set of inputs to a set of outputs. When the internal details of the system are not of particular interest, or are unknown, the mapping function between input and output spaces can be considered as an opaque box with one or more inputs entering and one or more outputs exiting the system (see Fig. 5.1).

When the inputs and outputs to the system are in digital form,[1] we would call this a *digital system*; when they are all analog, that is, not in digital form, we would call this an *analog system*. Current from a wall outlet represents an analog signal.

Systems can have both digital and analog inputs and outputs — we call these systems *hybrid* systems. Most natural systems would be considered analog and many human-built systems that involve computers or electronics would be considered digital, though some are not. For example, most toasters are purely analog. Human-built systems that do not involve electronics or computers are usually analog in nature. For example, a mechanical clock can be thought of as an analog system. Hybrid systems are very common too, for example, an automobile takes in and outputs all kinds of analog and digital signals to and from many devices. Even a lawnmower engine is hybrid because it involves mechanical parts and electrical parts (spark plugs) for ignition and combustion. In further discussions, we are not going to state whether a system is analog, digital or hybrid unless it matters.

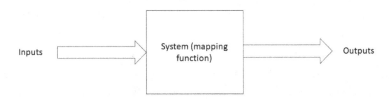

Fig. 5.1. A general system with inputs and outputs.

---

[1]For example, by letting high and low voltages represent logical 1s and 0s, respectively.

5.1.1 *Modeling Things as Systems*

Every real-world entity, whether organic or synthetic, can be modeled as a system. In computing systems, the inputs represent digital data from hardware devices or other software systems; the outputs represent control signals to switches, displays, or control signals to other devices. Vehicles, appliances and machines can be classified as analog, digital or hybrid systems, depending on the specifics. Living organisms can be modeled as analog systems — animals take in food, water, air, etc. and output heat, motion and waste. Plants take in light, minerals, gasses, etc. and have their own outputs.

Larger things can also be modeled as some kind of system, for example, factories, hospitals and cities. Collections of things can be modeled as systems, for example, armies, flocks of sheep and solar systems of planets. Weather, stock markets and the microclimate of a small valley are systems. Virtually, anything can be modeled as a system. In the broadest sense, any system f, that takes among its inputs time t, could be thought of as a dynamical system, though, more specifically, a dynamical system is one in which time is the primary determinant of behavior.

In any case, let's consider a simplified system f (the name for the opaque box) with one input x, which has certain properties that are defined by the set of all possible values of x, and one output y, which also has certain properties that are defined by the set of all possible values of y (see Fig. 5.2).

We would represent the transform of input x via mapping function f to output y by the equation:

$$y = f(x) \tag{5.1}$$

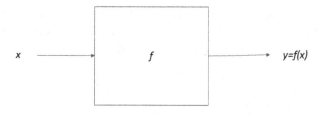

Fig. 5.2. A system with one input and one output.

In technical discussions, we would refer to these inputs as *stimuli* and the outputs as *responses* and we might refer to a single stimulus–response pair. In everyday parlance, we might refer to the inputs as *causes* and the outputs as *effects* and we might refer to a single cause-and-effect pair.

Mathematically, the x and y values are real numbers, but they could be integers, 1s and 0s or some other values depending on the type of system being modeled. The input x could also be one of a series of inputs presented either incrementally (e.g. as a set of digits) or continuously (e.g. as a current). There are usually constraints on any series of inputs, for example, that they cannot be of infinite value and that they must be presented over a finite time interval. But as usual, we omit the mathematical rigor for simplicity of presentation.

There are many real systems that can be modeled this way. For example, a shower system can be modeled as having one input (the stimulus) that is a clockwise or counterclockwise turn of a valve and one output, a flow of water at a certain temperature. Turn the valve to the right and the temperature of the water output from the faucet increases a certain amount (the response). Now, if you turn the valve to the left and the temperature decreases. Of course, there are limits on how much you can turn the valve in either direction or on the temperature of the water.

If we were to rigorously model this faucet system, we would need to address these limits as well as other inputs to the system, such as room temperature and atmospheric pressure. But for most purposes, the simple model of the faucet suffices. Compare this system to another one that involves turning something — a kaleidoscope.[2] Here, turning the dial in one direction or other leads to dramatic changes in the fractal-like images viewed through the eyepiece.

Many electrical and mechanical systems behave one way upon start up (or initialization) and then differently afterwards during regular operation. The period of regular operation after initialization may reach a *steady state*, that is, where very little changes. For example, when an automobile starts,

---

[2]The kaleidoscope was invented by the Scottish scientist David Brewster (1781–1868). Brewster was known primarily for his work in optics, but he made contributions to many other areas.

the engine goes through many different operations (and makes different sounds), then when it is idling (a steady state), which, in turn, is very different when driving. Many electronics systems go through various kinds of self-tests before entering regular operation (or some kind of waiting, steady state). Some systems can appear to be chaotic before entering normal operations or some steady state, for example, initializations can often appear that way. Some systems, for example, chaotic ones, never enter a steady state. The term *steady state* can also apply to natural systems as well. For example, the calm that occurs after a chaotic storm can be viewed as a kind of steady state.

We can also extend the system shown in Fig. 5.2 to include the modeling of a collection of systems connected end-to-end (i.e. in series), side-by-side (i.e. in parallel), with feedback and in other ways. However, we would have to be more mathematically rigorous to present these concepts, and we don't need them for further discussions. We encourage the reader, however, to investigate this aspect of systems theory (see Exercises).

5.1.2 *Linear Growth*

An important class of systems involve growth. Suppose the box in Fig. 5.2, f, now represents some kind of system that increases whatever you put in it due to some rule. The rule is usually a growth rate r that is based on a unit or interval of time. Lots of things can be modeled as growth systems. For example, the system could represent an investment that grows at rate of r percent per year, a plant that grows r millimeters per week, or it could be a magic box that increases the size of anything that you put in it at a rate r units every day. But for some theoretical systems, the growth could also be instantaneous. The following discussion applies to any linear growth system, but for simplicity, we'll cast it in financial terms.

Suppose you have P dollars in a bank savings account that pays interest at the rate r, compounded annually (meaning the interest is computed and paid once per year). The amount you have (the principal) will grow according to the schedule shown in Table 5.1.

As you add rows to Table 5.1, it starts to get messy in a hurry. But you can simplify things by noting that you can keep pulling out factors of

Table 5.1. Principal and interest for two years in a savings account.

Year	Balance
Now	P
Now+1	$P + rP$
Now+2	$(P + rP) + r(P + rP)$

Table 5.2. Simplified interest growth model.

Year	Balance
Now	P
Now+1	$P(1 + r)$
Now+2	$P(1 + r)^2$

$(1 + r)$ from each line. If you do that, the balances collapse to a simple pattern shown in Table 5.2.

If you follow this pattern out for Y years, you get the general formula for the future value of the money in the savings account (FV):

$$FV = P(1 + r)^Y$$

But this formula is only for interest that is compounded annually. Most interest is compounded monthly, weekly, daily, or even instantaneously. To get the general formula, we'll start out with interest compounded n times per year:

$$FV = P(1 + r/n)^Y n$$

where P is the starting principal and FV is the future value after n times of compounding in a year. When $n = 1$, it is just interest compounded annually. For $n = 12$, compounded monthly. For $n = 52$, compounded weekly and for $n = 356$, compounded daily (omitting the issues surrounding leap years).

To get to the instantaneous case, we have to do some calculus, and if you are unfamiliar with this, then skip to the end of the discussion. First, let's take the limit for an infinite number of infinitesimally small time intervals:

$$FV = \lim_{n \to \infty} P(1 + r/n)^{Yn}$$

We can simplify the right side by introducing a new variable, defining $m = n/r$, yielding

$$FV = \lim_{m \to \infty} P(1 + 1/m)^{Ymr}$$

pulling P, Y and r out of the limit since they are not affected yields:

$$FV = P\left[\lim_{m \to \infty} (1 + 1/m)^m \right]^{Yr}$$

The limit in the square brackets converges to the familiar number $e = 2.71828\ldots$ So, the formula becomes

$$FV = Pe^{Yr}$$

Does it really matter if the interest (growth) is compounded annually or otherwise, say, instantaneously? In fact, it makes a lot of difference. Suppose you have \$1,000 in a savings account that pays 2% interest per year compounded instantaneously. In 5 years, how much will your \$1,000 be worth?

$$FV_{\text{instant}} = 1,000e^{(5)(.02)} = 1,105.17$$

Compare this to the same \$1,000 in an account saved at 2% compounded annually.

$$FV_{\text{annual}} = 1,000(1 + .02)^5 = 1,104.08$$

There really is a difference.[3]

So, it is really important to understand how to model growth (in anything, not just interest in a bank account). Moreover, you can see that we had to simplify things considerably in this discussion, and even so, we ended up

---

[3]We hope this discussion helps you see the benefits of long-term investing and saving. Conversely, just as interest paid to you compounds exponentially, so does interest that you owe on loans, credit card debt and so on. Saving, investing and reducing debt are important parts of a strategy to financial independence.

with a nonlinear (exponential) model for growth. The moral of this story is that complexity can emerge in even simple systems once you get into the details.

5.1.3 *Response Time*

As previously noted, except in a purely theoretical system, there is a delay between the presentation of all inputs to a system, the realization of any required behavior and the appearance of all outputs. We define this delay as the *response time*. How fast and punctual the response time needs to be depends on the characteristics and purpose of the specific system.

For example, the fastest computer circuit switching delays can be on the order of billionths of a second, while it can take a minute or more to open a drawbridge spanning a river channel. Signals sent to space vehicles far away can take days, months or years to reach their destination and it can take decades or even centuries to note the shift in tectonic plates or the movement of glaciers. There is even a noticeable delay in stimuli to your sensory organs and their receipt by the brain — it ranges from about 15–30 milliseconds to process an image captured by the eye.[4]

The fact that all real systems have some time requirement for processing leads to the observation that every real thing can be modeled as a dynamical system.[5]

5.1.4 *Control Systems*

Often, a human-built system is controlling another system.[6] That is, the outputs from a system are the inputs to a *system under control* and there

---

[4]Since there is a response time delay for all senses from stimulus to processing by the brain, it is could be said that we are always living in the past.

[5]What about a rock? If it's a theoretical rock, then probably, no. But real rocks occur in nature and are subjected to the forces of gravity, pressure, wind, water, and so on. They can be moved by earthquakes, tornadoes, floods and humans or animals, and therefore, their position is a function of time. So, yes, a rock can be modeled dynamically. In fact, geologists use dynamical systems modeling techniques all the time.

[6]It is also possible to model the interactions of living things or natural phenomena as control systems.

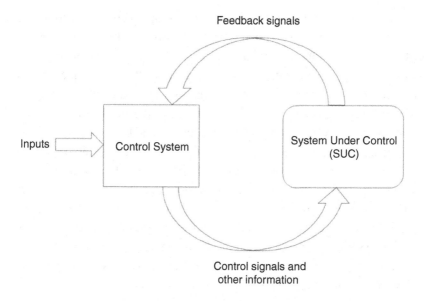

Fig. 5.3.   Control system block diagram.

may be outputs from the system under control that are *feedback* inputs to the controlling system (see Fig. 5.3).

For example, in the shower, you are the controlling system, and the shower is the one under control. There is a feedback loop between you and the shower — you feel the water temperature, decide to change the temperature, turn the valve, and then wait for the water temperature to change. In some cases, there can be a long delay, and this can lead to what appears to be chaotic behavior. For instance, you get in the shower and the water is cold. You turn the valve to the right to make it warmer, but there is no perceivable change. You turn the valve to the right again. Still no change. You turn the valve to the right again, then, wow, all of a sudden the water is scalding hot. Now, you turn the valve to the left to cool the water. No change. You turn the dial left to cool it again. Still no change. You turn the valve left again, then, bam! the water is freezing again. The problem here is that time-delayed control systems can appear to be chaotic when there is a significant lapse in time between stimulus (turning the valve) and response (change in temperature).

The same kind of behavior can be found in supply chains. Suppose there is a sudden spike in demand in a certain product, say, kale.[7] The stores are suddenly stripped of kale by eager buyers. The store immediately orders extra kale from the produce distributor, which itself sees a sudden demand in orders for kale from many other stores. The distributor places additional orders with the growers — of course, the growers need time to increase the supply. Meanwhile, demand for kale, which had been increasing begins to subside as disappointed consumers find alternatives to kale. Finally, the distributor receives enough kale to fulfill the backlog from the stores. The kale is delivered to the stores which are now overstocked with kale since consumers are no longer as interested in the product. So, now the stores stop ordering kale for a while because they are overstocked. Of course, as the kale in the stores is bought or disposed as trash as it rots, a shortage of kale is again created, which will take time to alleviate because of supply chain delays. Consumers see this whipsaw effect as "kale chaos," but we know it is just a manifestation of a substantial stimulus–response delay time.

In some cases, this "significant" delay in stimulus–response can be quite small. For example, we mentioned that there is a 15-30 ms delay between when your eye captures an image and when your brain processes it. In certain simulators and games (e.g. involving fast flight), if this delay is not represented correctly, or changes too much, it can lead to nausea of the operator or player. If the problem is caused by the system responding too quickly to a stimuli, a delay needs to be built.

By the way, this feedback cycle for a human and another system is called an OODA loop. OODA[8] is an acronym that stands for "observe," "orient," "decide" and "act." That is, an individual has to be stimulated by the presence of the threat (usually visually, though it could be by any of the other senses), then orient itself to the nature of the potential threat, decide how to respond, and then act accordingly.

---

[7]The cause for this spike could be anything. Maybe a recent report on a popular news program claimed that kale cures cancer. Or maybe the kale growers consortium unleashed a major marketing campaign (but forgot to inform the growers).

[8]The OODA concept was developed in the 1950s by US Air Force Colonel John Boyd to describe and study aerial combat techniques.

The OODA loop helps explain why it takes time for someone to react to a threat and how an attacker (or defender) can disrupt the observe and orient phases of the OODA loop by overloading sensory inputs or through evasive actions. Moving in a way that is counterintuitive or deceptive can also short-circuit the decision-making stage of the OODA loop.

The OODA loop is used in many robotics applications and it is also used as a cybersecurity strategy because it is a concept based on having a clearer focus to improve our ability to be adaptable to unexpected circumstances.

5.1.5 *Deterministic Systems*

Control of any system is maintained when given the current state and a set of inputs, the next state of the system, is predictable. In other words, the goal is to anticipate how a system will behave in all possible circumstances. That is, a system is *deterministic* if for each possible state and each set of inputs, a unique set of outputs and next state of the system can be determined.

Event determinism means the next states and outputs of a system are known for each set of inputs that trigger events. Thus, a system that is deterministic is also event deterministic. Although it would be difficult for a system to be deterministic only for those inputs that trigger events, this is plausible, and so, event determinism may not imply overall determinism.

It is interesting to note that while it is a significant challenge to design systems that are completely event deterministic, and it is possible to inadvertently end up with a system that is non-deterministic, it is very hard to design systems that are deliberately non-deterministic. This situation arises from the utmost difficulties in designing perfect random number generators. Such deliberately non-deterministic systems would be desirable, for example, in casino gaming machines or password generation algorithms.

Finally, if in a deterministic system the response time for each set of outputs is known, then the system also exhibits *temporal determinism*.

A side benefit of designing deterministic systems is that guarantees can be given that the system will be able to respond at any time and, in the case of temporally deterministic systems, when they will respond.

5.1.6 *Time-Variant and Time-Invariant Systems*

We noted that, strictly speaking, all real systems are affected by time, that is, they are a function of time. But when time is not the primary variable, or, in the case of a theoretical system, not a variable at all, we call these systems *time-invariant*.

Many electrical and electronic systems are time-invariant. For example, toasters, television sets and lights. Mechanical devices, such as hand-powered lawnmowers or block and tackle systems are usually time-invariant. You expect all of these systems to work the same every time you use them during their effective lifetime, regardless of the time of day or the day of use. Of course, those components can change performance characteristics after long periods of operation (or fail early because of some manufacturing defect), but for the most part, we consider these systems to be time-invariant.

If time significantly affects a system's behavior, then we call these systems *time-variant*. Clearly, biological and ecological systems where there are relatively short life spans of organisms are a good example class of time-variant systems. Certainly, the characteristics and behaviors of the human body, an animal or plant tend to change significantly over time. Weather phenomenon are time-variant systems as well — tides and waves ebb and flow, storms simmer, rise up and subside, tornadoes and hurricanes change behavior over time before they disappear. Of course, certain electrical and mechanical devices are designed specifically to be time-variant — clocks and timers, random number generators, lighting, irrigation systems, and many other types of systems.

5.1.7 *Linear Systems*

A *linear system* is a system where the behavior exhibits certain special characteristics. Intuitively, a linear system is one in which its responses are proportional and additive to any inputs to it.

Mathematically, if a and b are real numbers and for all x the input space, then

$$f(a \cdot x_1 + b \cdot x_2) = a \cdot f(x_1) + b \cdot f(x_2) \qquad (5.2)$$

holds for all possibilities of a and b, then f is a linear system. Time-invariant systems are linear with respect to the variable of time.

For example, consider the function $f(x) = 10x$, which is linear. Let's see why. Suppose a and b are arbitrary real numbers. Then

$$f(ax_1 + bx_2) = 10(ax_1 + bx_2) = 10ax_1 + 10bx_2$$

Now, $af(x_1) = 10ax_1$ and $bf(x_2) = 10bx_2$. So,

$$af(x1) + bf(x_2) = 10ax_1 + 10bx_2 = f(\dot{a}x_1 + bx_2)$$

and therefore f is linear.

From a systems standpoint, linearity is not exactly what you might expect. Consider the function $f(x) = 5x + 2$, which is, of course, a straight line in slope-intercept form. But does $f(x)$ represent a linear system? Let's see. Suppose a and b are arbitrary real numbers. Then

$$f(ax_1 + bx_2) = 5(ax_1 + bx_2) + 2 = 5ax_1 + 5bx_2 + 2$$

Now,

$$af(x_1) = a(5x_1 + 2) = 5ax_1 + 2a \text{ and } bf(x_2) = b(5x_2 + 2b) = 5bx_2 + 2b$$

So,

$$af(x1) + bf(x_2) = 5ax_1 + 5bx_2 + 2 + 5ax_1 + 5bx_2 + 2$$
$$= 10ax_1 + 10bx_2 + 4$$
$$\neq 5ax_1 + 5bx_2 + 2 \tag{5.3}$$

So, f is not linear. So, even though this f represents a straight line, because it has a non-zero y intercept, from a systems theory perspective, it is non-linear. This is a bit confusing, but to repeat, just because something can be approximated or exactly represented by a straight line, it does not mean it is linear from a systems theory perspective.

Is the shower system linear? We would say yes. Intuitively, if you turn the valve twice to the right, then once to the left, you should get the same temperature as if you turned the valve only once to the right. Of course, if you don't rotate the valve exactly the same amount each time, things won't turn out quite right. But from a theoretical standpoint, the shower system behaves linearly. You can model the valve as a function to show that it obeys the property of linearity. Let's assign t as the real variable of temperature and r as a real variable, the r represents the rotation of the valve, with positive r meaning rotation to the right for hotter water and negative r representing rotation of the valve to the left for a colder temperature (with $r = 0$ meaning no rotation and the same temperature). Then the temperature at the faucet is some linear function of r, that is,

$$t = f(r)$$

It is easy to show that this is a linear function. Of course, we have to bound the variable r by the physical limits of the valve's rotation range, but aside from that, it obeys the linearity equation. A takeaway here is that it takes lots of assumptions (and mathematical precision, which we have omitted here) to model real-world system as being actually linear.

When a system is linear, it is possible to use the responses to a small set of inputs to predict the response to any possible input. This aspect is very valuable to scientists. Real-world linear systems tend to be those that can be modeled or approximated by a straight line with the input variable as the x-coordinate and the output variable as the y-coordinate and with the y-intercept being zero.

Some linear systems in nature include (simplified) growth models and many mechanical and electrical systems. For example, *Hooke's Law*[9] states that for small deformations, the stress and strain on a spring are proportional to each other, that is, $F(x) = -kx$ where F is the force, x is the extension length and k is the constant of proportionality known as the *spring constant*.

Strictly speaking, real-world linear systems are not that common; they are idealized models purely for the simplicity of theoretic analysis and design. However, the system is essentially linear when the magnitude of the signals in a control system are limited to a range in which the linear characteristics exist.

---

[9]Robert Hooke (1635–1703) was an English scientist and architect who among many important achievements in physics and mathematics was the first person to visualize a microorganism using a microscope.

5.1.8 *Nonlinear Systems*

A *nonlinear system* is a system that is not linear, that is, it has a system response function that does not obey equation (5.2). For example, consider a system that has a response f to input r that can be modeled by the function

$$f(r) = 4/3\pi r^3 \tag{5.4}$$

You might recognize this as the formula for the volume of a sphere of radius r. In any case, a system that has this response function is nonlinear, and we can show this easily by a counterexample, that is,

$$f(4 \cdot r + 3 \cdot r) \neq 4 \cdot f(r) + 3 \cdot f(r)$$

You can work out the details of this yourself (see Exercise 5.3).

Intuitively, nonlinear systems and nonlinear phenomena are those where "the sum of the parts does not equal whole." For example, suppose you study for two years to be an engineer. You stop and decide to spend two years to study to be a nurse. That's four years of study. But are the outcomes of those four years of study the same as either four years of study as an engineer or as a nurse? Probably not. In this nonlinear phenomenon, the payoff for focusing for four years on one profession is likely going to be higher than focusing two years each studying different professions.

From a systems perspective, nonlinear systems respond differently if you break up a set of inputs into parts instead of providing all of those inputs at the same time. For example, the time-varying heat loss $P(t)$ in a resistor of r ohms in which a time-varying current $i(t)$ is flowing is given by

$$P(t) = i(t)^2 r$$

So, if you were to apply a current of 5 amperes for 2 seconds and then apply another current of 5 amperes for 2 seconds, you would get a very different heat loss profile than if you were to apply a current of 10 amperes for 1 second. Even if you are not familiar with electricity or electronics, you can see this behavior is nonlinear since

$$5^2 r + 5^2 r = 50r \neq 10^2 r = 100r$$

So, whenever the current is non-zero, the heat loss is different in the two scenarios.

We already noted that systems that can be modeled or approximated by a straight line are only linear if the y-intercept is zero. Trigonometric and exponential operators, which are found in many fractals, are also nonlinear.

Nonlinear behavior can appear to be chaotic, but nonlinearity does not necessarily mean that the system is chaotic. A system can be nonlinear and not chaotic (or it could be nonlinear and chaotic). Conversely, it is hard to conceive of a chaotic system that is linear and there are likely none. Quantum physicists, however, sometimes debate the possible existence of chaotic linear systems.

5.1.9 *Summary of System Types*

In any case, to summarize, we can have linear, time-invariant systems, linear time-variant systems, nonlinear time-invariant systems and nonlinear time-variant systems (see Exercise 5.1). Modeling real-world phenomenon in any of these ways help scientists and engineers to understand these systems and make predictions about them. But when systems are chaotic, it is very difficult or impossible to understand them well enough to make predictions about them.

Hopefully, though, you too will begin to observe natural and human-made phenomena and look for these systems characteristics. Just as scientists gain insight into systems behavior by classifying its type, so can you.

5.2 Systems Thinking

Systems thinking is concerned with making sense of complexity by looking at the bigger picture of systems and their interrelationships with other systems and the environment. Looking back at Figs. 5.1 and 5.3, we see that these are closed systems, that is, nothing else is considered except inputs and outputs. But is this a realistic assumption? What about environmental factors? What about unseen or accounted for systems that might interact with these systems? It is unrealistic to assume that any system does not interact with its environment in some way (see Fig. 5.4). Systems thinking would have us believe that there is more than this abstract notion of a system and its interaction with the real world.

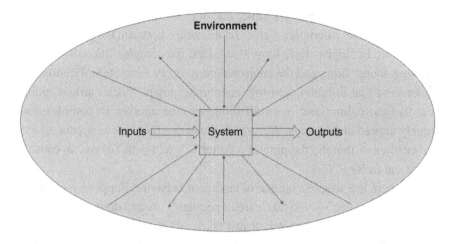

Fig. 5.4. A real system interacts with the environment in obvious and subtle ways.

We know that if the system is chaotic, even the tiniest perturbation from some environmental nuance can have significant effects. Whether the system is open or closed, whether we understand the nature, magnitude, and direction of the inter-relationships of a system's parts or not, we need to decouple those parts and examine them individually.

There is an old adage: "The whole is greater than the sum of its parts," which is not entirely correct, since the whole could actually be less than the sum of its parts. But what is certain is that the whole is not equivalent to the sum of its parts. The importance of each component of a system is tied to its relationship to the whole and the essential properties of a system are properties of the whole. Let's look at three important aspects of systems thinking: emergent behavior, second-order effects and uncertainty.

5.2.1 *Emergent Behavior*

By looking at just one system component in isolation, we may not have a realistic picture of its importance. An *emergent behavior* or *emergent property* can appear when a number of simple entities operate in an environment, forming more complex behaviors as a collective.

In nature, no system is simple, and we do not find isolated building blocks, but rather a complex web of relationships between the parts of a unified whole. Ecologists have long recognized the complex interconnections between fauna, flora, and the environment in every ecosystem. Economists understand that inflation, housing, economic growth, stock markets, political factors, culture and news all influence one another in complex and largely unpredictable ways. We already know that there are emergent effects on weather — that the flapping of a butterfly's wings in Tokyo can cause a hurricane in New York.

But here is a simple example of emergent behavior. Suppose you have a pet sitting business — you take care of peoples' cats and dogs at your home while their owners go to work or just take care of errands, travel, etc. You charge $25 per day per animal. People drop off their pets from 7 to 9 AM and must pick them up anytime before 6 PM when you business is "closed."

You love animals, but this is hard work, and on many days, some of the owners are very late picking up their cat or dog, which often means you can't take care of your own activities or eat dinner until 8 PM or later. To discourage this behavior, you begin instituting a late fee — for every hour or fraction thereof that the animal is picked up late, you charge an additional $5. But a strange thing happens — owners who were picking up their pets late leave them there longer and even more owners are picking up their pets late than before. Although you are making some extra money, this emergent behavior is not what you wanted — you are tired by the end of the day and you had hoped all owners would pick up their pets on time.

But why did this behavior emerge? There are several possibilities, but a likely one is the following. The owners that were picking up their pets late are happy to continue to do at the reasonable price you are charging, and now don't feel guilty in doing so. Many other owners had wanted to leave their pets longer but wanted to honor the pickup deadline. But now that you are offering a reasonably priced extension to the pickup time, they feel free to leave their pets longer. So, what could you do to eliminate this emergent behavior? There is a lot more to the economics of this little business, but one easy answer might be to charge much more for each hour that the pet is left late. But this too might create an emergent behavior, for example, some of your customers might see this new policy as overcharging and move their business to another sitter.

The possibility of emergent behaviors can make even an apparently simple system's problem much more complex. In some cases, the emergent behavior can appear chaotic (because of sensitive dependence on small changes). Therefore, it is very important when making changes to systems (of whatever kind — business, electronic, mechanical, ecological, educational, etc.), the possibilities of emergent behavior and their undesirable outcomes (or even desirable ones) are monitored and addressed.

5.2.2 *Second-Order Effects*

Second-order effects (sometimes called side effects or unintended consequences) refer to the observation that in systems every action has a consequence. This consequence, if notable and impactful, is called a *second-order effect*. For example, in the shower control system, making the water too hot can lead to a burn — this is a second-order effect of turning the valve.

Second-order effects illustrate another law of systems thinking, that is, the cause and the effect are usually not closely related in time and/or space. The delay between turning the shower valve and the change of water temperature clearly is an example of this law.

Second-order effects may have additional, higher order effects, in essence, creating a chain reaction. For example, turning the valve causes a person to be burned and react in such that they bang their head on the shower door, which causes the door to shatter. When they step out of the shower, they cut their foot on the broken glass and so forth. Of course, many serious car accidents have been caused by a chain reaction of events — a driver looks down at their phone, causing them to drive through a red light, which initiates a series of crashes of cars at the intersection and beyond. Additional side effects include injuries, damage to property and an extended traffic jam (which causes drivers caught up in it to miss appointments).

Other examples of higher order effects include the assassination of Archduke Ferdinand leading to a chain reaction of events that led to WWI. Also, remember Richard the III quotation from Chapter 1. Higher order effects can appear as chaos (e.g. the chain reaction car accident described above) or become fractal-like (e.g. see Section 5.4).

5.3 Uncertainty

Uncertainty is as much a part of science as it is in life. Indeed, the measurement of physical data is affected by instrumentation in which accuracy is constantly in question. In addition, philosophers and psychologists will argue that even if our instrumentation were perfect, our perception of the readings is subject to the distortions of the limitations and frailties of our mind and body. Let's look at several different kinds of uncertainty and, later, ways to deal with uncertainty.

5.3.1 *State Uncertainty*

If we view the real-world system as a state-based transformation of a set of inputs and current state into a new state and set of outputs, each of these elements can incorporate uncertainty. For example, uncertainty can be found in any one of the inputs to the system. Similarly, the state of the system might be uncertain at any time, resulting in loss of control. The transition from one state to another can also be non-deterministic, i.e. uncertain. Moreover, the outputs to the external environment are not always predictable in poorly designed systems. Finally, the real-world system must interact with an environment that could be uncertain.

Uncertainty in time arises from the fact that the time is not necessarily deterministic for the system to transition from an input set and current state to the output set and new state. Here, we are faced with a dilemma: solving many problems with inherent uncertainty are themselves *NP-hard* or *NP-complete problems*.[10]

---

[10]Informally, NP-hard problems are computational problems that can only be attacked with brute force approaches — there is no known polynomial-time algorithm that solves any single problem in the class. The class of NP-complete problems has the following properties: they are NP-hard and any particular problem in the class implies that every other NP-complete problem can be solved with a polynomial-time algorithm. Many games contain aspects of combinatorial or graph problems that are known to be NP-hard or NP-complete. Even a seemingly straightforward problem, such as packing a backpack with food that has limits on weight and volume with constraints on calories, fat, protein, etc., is an NP-complete problem.

Another kind of state uncertainty can arise in even the simplest of systems. A *bistable system* is a device or a system which can exist in only one of the two possible states (and not both). For example, a simple light switch (not a dimmer switch) is bistable — the switch is in either the on or off position corresponding to the light being on or off. Digital computers rely upon billions of tiny bistable logic switches that are either in a "0" or "1" state. Many mechanical devices are bistable, for example, a mechanical door lock that is in either the locked or unlocked position. But certain bistable devices can experience situations in which the state is unsettled. Contact (or switch) bounce is a physical phenomenon that occurs because it is practically impossible to build an electromechanical switch that could change its state instantaneously without any contact oscillation (bouncing). Phantom state changes triggered by various buttons, relays, and switches all exhibit this undesired phenomenon, which creates a form of uncertainty that can appear to be chaotic. Furthermore, some systems, which appear to be bistable, are inherently not. For example, a door, which can be open or closed, can really be in any one of an infinite number of positions in between, creating another kind of uncertainty.

5.3.2 *Heisenberg Uncertainty*

In 1927, Werner Heisenberg[11] postulated his famous uncertainty principle, which states that the precise position and momentum of a particle cannot be known simultaneously; therefore, trying to be more certain about one comes at the expense of increased uncertainty in the other. Heisenberg uncertainty was noted in Chapter 3 as a metaphor for observing weather.

Real-world application of Heisenberg uncertainty is metaphorical since it is impossible to interact directly with a single electron. Instead, we can think of uncertainty in real-world systems existing along three broad dimensions: time, space and behavior. Like Heisenberg uncertainty, trying to improve the certainty in one system aspect comes at the cost of uncertainty in another. For example, trying to bring uncertain behavior under control almost always requires additional time or space (for more hardware

---

[11]Werner Heisenberg (1901–1976) was a German theoretical physicist and one of the pioneers of quantum mechanics.

or electronics). On the other hand, "cheating" on timeliness, perhaps by prematurely terminating some calculation or action, can lead to uncertain behavior. Finally, there is the fundamental time–space tradeoff; removing uncertainty in time usually increases uncertainty of physical requirements (space) and vice versa.

5.3.3 *Uncertainty and Chaos*

Chaotic systems are those in which small changes in inputs lead to radically changed behavior and outputs. Chaotic systems present significant challenges to the systems engineer. Complicating the situation is the fact that corrupted outputs from the real-world system to the system under control can cause a stable system under control to appear to be chaotic.

For example, imagine trying to balance a long stick with the end placed in the palm of your outstretched hand. This is challenging, even for a few moments. Now, imagine trying to balance the same stick in your palm, but now there is a brick balanced on the top end of the stick. This is much more challenging and dangerous if the brick were to fall. Now, imagine trying to maintain this precarious balance with your eyes closed — even more challenging. We can add other environmental factors such as the wind blowing or other people even trying to push you to make it nearly impossible to maintain the balance. You might not call this situation chaotic, but you also probably wouldn't consider achieving a steady state of balance to be easy. We have only made a short list of complicating problems — there are many other uncertainties that could make this situation appear very chaotic.

It turns out that building a system that can automatically control a moving platform with an end-balanced pole and a movable weight on top is very difficult. In fact, this situation is a close approximation of the large boom lift machines that you may have seen at building construction sites.

5.3.4 *Dealing with Uncertainty*

Nature deals with uncertainty in many different ways. It is said that "nature abhors a vacuum," meaning, in one sense, that uncertainty is always resolved in some way. For example, when the top-line predator in an ecosystem is led to extinction or removed, it is unclear how the ecosystem will resolve the void. But in time, the situation resolves itself, for example, a new top-line predator (or predators) emerge to fill the vacuum. Or, worse, the ecosystem becomes so destabilized that overpopulation ensues. You can play with the wolf–caribou model discussed in Chapter 3 to see what would happen to the caribou population if the death rate for wolves was very high.

Uncertainty has and always will be a vexing problem for systems engineers, information scientists, scholars and ordinary citizens. New techniques are created all the time for dealing with uncertainty, but really no one has all the answers with respect to this problem. But there are several mathematical frameworks for modeling uncertainty. Typical approaches include expert systems, probabilistic reasoning, fuzzy sets and logic, neural networks and rough sets. Let's discuss some of these briefly. Further study is offered as well (see Exercise 5.14).

5.3.4.1 *Expert Systems*

An *expert system* is a form of artificial intelligence in which a software program imitates a human expert in some restricted domain of knowledge. A good example of an expert system, which you have likely encountered, is the recommender system for an online shopping site. Here, your past purchasing history and other information about you and people like you are used to make recommendations to purchase products that you are likely to buy.

As another example, consider the online driving maps and GPS information systems that provide drivers with real-time information about traffic, accidents, closed roads and so forth. Weather apps and news stations also provide information about the likelihood of rain, wind, fog, etc. which can affect driving conditions. The human driver then uses this information to informally calculate the probability of future traffic jams, choke points and detours. Accordingly, the driver then may alter their route and speed. In the same way, an autonomous driving system could use this information to make similar decisions and take similar actions.

In other kinds of expert systems, pure knowledge-based and hybrid knowledge representation help to automate the processing of uncertain information. For example, in real-time databases, they can deal with inconsistent, missing or corrupted data.

But the behavior of expert systems can sometimes appear confusing or chaotic to humans. This is why a desirable property of expert systems is that they are *explainable*. This property means that the decisions and actions made by an expert system can be easily explained and are understandable to humans.

5.3.4.2 *Probabilistic Reasoning*

Traditional probability theory can also be used to describe the uncertainty of the interconnections between hardware and software components. Using probability theory, an event (or experiment) E is characterized by one or more outcomes $o_1 \ldots o_n$. Each outcome o_i is assigned a probability on the continuum $[0, 1]$. A certain event has a probability of 1 and an impossible outcome has a probability of 0. Thus, the probability of an uncertain event $0 < o_i < 1$. The more certain the event is to having a outcome o_i, the closer to 1 that probability is. If less certain, then the probability is closer to 0. However, the sum of the probabilities of all outcomes must equal 0.

As a simple example, consider the event of flipping a coin. Here, there are only two outcomes: heads or tails.[12] If the coin is fair, then the probability of each outcome, heads or tails, is 0.5. However, if the coin is weighted such that it always lands heads, then the probability of getting a head is 1 and the probability of getting a tail is 0. But if we told you that the coin was weighted such that it usually lands heads (but not always), now some uncertainty is introduced. What is the probability of getting a head for this lightly weighted coin — 0.3? 0.9? We can only find out through many repeated experiments of flipping the coin. Thus, we see that there is uncertainty in this event.

As an example, consider the autonomous driving system just described. Based on the weather, traffic and other information, the system informally calculates the probability of future traffic jams, choke points and detours and takes the appropriate actions — just as a good human driver would.

---

[12]You might argue that there is a third possible event — the coin landing on its side. This event might be possible but extremely unlikely — so unlikely that the probability of this event is nearly 0, so near that we can treat it as 0.

In modeling systems, the probabilities of the outcomes of events can be propagated through the many state changes of the system over time. Then probabilities can be assigned to failure outcomes. These probabilities can also be used in various reliability analyses to deduce what happened to a failed system and what likely caused any failure.

5.3.4.3 *Fuzzy Values, Fuzzy Logic and Fuzzy Sets*

Fuzzy valuation provides a mechanism for dealing with the classification of some property that is not "crisply" defined as having or not having that property. For example, is a certain knife sharp or dull? Actually, the property of sharpness can be viewed on a continuum from "extremely sharp" to "very dull" with all levels of sharpness/dullness in between. Other fuzzy properties might include large or not large (small) for some organism, fast or not fast (slow) for some process and clear or not clear (opaque) for some object. All of these properties (and their compliments) are well modeled with fuzzy values.

Traditionally, uncertainty is modeled in fuzzy valuation by providing a fuzzy valuation function that measures possession of a property along the unit interval (with the number 0 meaning does not have and 1 meaning completely has the property). For example, the dullest knife would have a fuzzy value of 0 and the sharpest knife a fuzzy value of 1, and a marginally sharp knife a fuzzy value of 0.5. Likewise, the largest animal would have a fuzzy size value of 1 and the smallest animal a fuzzy value of 0. Some medium-sized animals might have a fuzzy size value of 0.5 or 0.6 Special operations on these fuzzy-valued functions need to be defined in order to combine and process information that is fuzzy-valued (we encourage you to investigate these).

Fuzzy logic is similar to boolean logic, but unlike boolean logic, fuzzy logic variables may be any real number between 0 and 1, so the logical operations and set operations have to be handled differently. That is, for fuzzy values x and y, the boolean AND equivalent is $\min(x,y)$ and the OR equivalent is $\max(x, y)$, and the boolean not operation equivalent for x is $1 - x$.

Fuzzy sets involve the notion of something belonging to or not belonging to a set that is defined in some way. However, partial membership is permitted through the use of fuzzy valuation. In fuzzy sets, the traditional relations of set containment, subset, and so on have to be redefined with fuzzy values. The set operations of union, intersection and compliment also need to be redefined.

We could use these fuzzy values, fuzzy logic, and fuzzy sets in order to model and control situations where the inputs and outputs are not crisply defined and where there is uncertainty. But we would need to extend the mathematics significantly. We encourage the interested reader to investigate these concepts further.

5.3.4.4　*Neural Networks*

A neural network is type of information processing structure intended to imitate the thinking and behavior of human (and other higher order animal) brains. A neural network consists of neurons interconnected via unidirectional signal channels called connections. Neurons have a local memory and execute localized information processing operations.

Each neuron has a single output connection that can be of any mathematical operation desired. The information processing within each processing element can also be defined arbitrarily and depends only on the current values of the input signals arriving at the processing element and on values stored in the processing element's local memory. These values are adjusted via weighting factors, which can be adjusted over time as the neural network "learns." This learning power is the basis for the neural network's ability to deal with uncertainty.

5.3.4.5　*Rough Sets*

Rough sets use the theory of unifying a collection of differently bounded sets to describe the greater whole. For example, when dealing with images, a camera with higher resolution can capture more details of an object than a camera with a lower resolution. However, the lower resolution camera captures a wider field of the image. By mathematically combining images from many high- and low-resolution pictures, a more complete and accurate image is assembled.

5.3.4.6 *Multiple Sources of Uncertainty*

When dealing with the uncertainty introduced by multiple conflicting information sources, other mathematical techniques must be used. For example, in determining its position, a spacecraft may use GPS data, information from inertial navigation instruments, and star mapping. All three sources may provide a (slightly) different indication of position and this must be reconciled. Each form of measurement also has different error properties and precision.

In many cases, we may perform some kind of data fusion and uncertainty resolution when we get conflicting information about the same event from different news sources. In other cases, we filter this information from our own perspective to form a conclusion about what really happened. A technique called *Kalman Filtering* may be used to deal with these kinds of situations. A general class of solutions that deals with arbitrating the different information coming from a set of sensors that are measuring the same source of data is called *sensor fusion*. Kalman filtering and most sensor fusion algorithms are relatively complex and involve higher order mathematics than we can cover here.

5.4 Swarm Behavior

We turn again to nature to close our discussion on dynamical systems and systems theory by introducing a dynamical system with fractal-like emergent behavior — swarms. You probably have seen the beautiful but mystifying images of large numbers of birds flocking, fish schooling, insects (such as bees and ants) swarming, and even interesting moving patterns in crowds.

But while these are sophisticated living creatures, none of these have very high levels of intelligence — certainly not fish and insects (we'll let you wonder about crowds). Even birds, while intelligent enough as individuals to be trained to carry messages, do tricks, hunt for their owner and even speak and sing human songs, seem incapable of the coordination to produce the beautiful and changing images that they can produce *en masse* (see Fig. 5.5).

Fig. 5.5. Honeybee swarm near Wilmington, Delaware, courtesy of Small Wonder Honey, https://www.facebook.com/smallwonderhoney.

Herds of buffalo, antelope or sheep certainly comprise more intelligent animals, but it is not known that their mass movements are purposefully coordinated.

So, how do these animals create complex group maneuvers? Scientists have determined that when each individual follows a few simple rules, the collective emergent behavior becomes much more complex.

Returning to the behavior of crowds, we know they can inadvertently move as if highly coordinated. We sometimes even refer to human group behavior that follows a pattern (aggregate simple behavior manifesting as complex group behavior) as a "hive mind." Hive mind behavior can sometimes be found in driving situations, on social networks and in voting patterns. For example, if on a busy highway every driver swerves left and then right to avoid pothole, it may appear to be a carefully choreographed maneuver involving dozens of cars. But as with the swarming of less sophisticated creatures, the complex movement of crowds and hive mind behavior are related to emergent properties of massed simple behaviors, not some overarching control mechanism.

5.4.1 Simulating a Swarm

The simplest mathematical models of animal swarms generally represent individual animals as following three rules:

(1) Move in the same direction as your neighbor.
(2) Remain close to your neighbor.
(3) Avoid collisions with your neighbor.

These rules look rather like the rules for cellular automata, and sometimes cellular automata can behave like swarms. In any case, we can use these rules (or similar versions) to produce simulated swarms and flocks.

Algorithms to simulate swarms are relatively simple. They recognize that each individual in the swarm (flock, school, etc.) moves in approximately the same way and at the same speed. An example of the rules of swarm behavior are found in the work of Antelmi (2019):

(1) Steer in order to avoid crowding local swarm-mates.
(2) Steer toward the average heading of local swarm-mates.
(3) Steer to move toward the average position of local swarm-mates.

So, if each individual moves in this way, we get the swarming behavior.

There are many implementations of swarm behavior (including some in Rust). A nice swarm simulator can be found at Wolfram Research's demonstrations site: https://demonstrations.wolfram.com/BoidsSimulated FlockingBehavior. Figure 5.6 shows a snapshot for the simulation at one instant in time.

Doesn't the "swarm" configuration somewhat resemble the real swarm in Fig. 5.5?

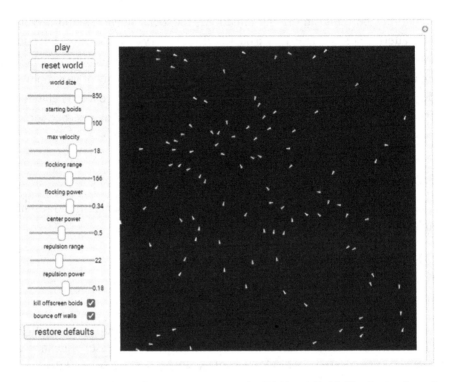

Fig. 5.6. Screen capture of an artificial swarm using Wolfram's Boids Simulator at https://demonstrations.wolfram.com/BoidsSimulatedFlockingBehavior. ©Wolfram Research.

Wolfram's simulator is based on Reynold's 1987 "Boids" flocks and swarm simulator model (Reynolds, 1987). The simulator allows for tweaking various parameters, such as flocking range, flocking power and repulsion range. In the left part of Fig. 5.6, you can see these controls. Try playing with these to create different and interesting swarming behavior.

5.4.2 *Applications of Simulated Swarms*

Swarm simulations have been used to create complex interactive systems or simulate crowds. They have also been used in many movies to simulate swarms, flocks, herds and crowds. For example, *The Lord of the Rings* film trilogy used swarm-like simulations to render battle scenes.

Small robots can also be implemented to swarm for various applications. Swarm technology is particularly attractive because it is relatively inexpensive to implement.

For example, the US space organization, NASA, is studying the use of satellite swarms to create new mission capabilities and complexities. One mission could be weather and climate data collection, another could be monitoring pollution, and yet another could be following a natural disaster, such as a wildfire. The conventional approach of controlling individual satellites does not scale for large numbers of satellites, so a swarm must operate as a unit, responding to high-level commands and constraints.

Swarming robots can also be used for defensive purposes, for example, to deny restricted airspace, protect maritime and subsurface platforms, and operate across large land, sea and air spaces inexpensively.

5.4.3 *Swarm Computing*

Computer scientists have used versions of the swarm rules as a way to solve difficult computing problems. This so-called *swarm computing* is a biology-inspired approach to decentralized systems composed of relatively simple agents that can self-organize to solve complex problems together through direct or indirect interactions. It emphasizes a special form of swarm intelligence, a behavioral metaphor for solving distributed problems on the basis of the principles underlying the behavior of natural multiagent systems, such as ant colonies and bird flocks.

Swarm intelligence provides a new behavioral model for multiagent systems stemming from local interactions between individuals with simple rule sets and no global knowledge. Coherence and cooperation emerge from a global viewpoint without any active push for it at the individual level. There are several key aspects of swarm intelligence:

- *Flexibility*: The swarm must respond to internal disturbances and external challenges.
- *Robustness*: Tasks are completed even if some individuals fail.
- *Scalability*: The swarm will range from a few individuals to millions.
- *Decentralization*: There is no central control in the swarm.
- *Self-organization*: The solutions are emergent rather than pre-defined.

The latter aspect allows the swarm to deal with uncertainty by adapting to unknown situations.

Swarm computing applications include optimization algorithms, communications networks and robotics.

5.5   Brief History of Dynamical Systems and Systems Theory

A complete history of dynamical systems and general systems theory would read like a who's who of mathematics and physics. More concisely, the studies of gravity and the laws of motion by Isaac Newton[13] and his contemporaries can certainly be listed as an important point in the study of dynamical systems. Poincaré's work on celestial mechanics in 1899 marked a significant achievement in describing dynamical systems of the solar system. Poincaré[14] is considered one of the fathers of chaos theory.

5.5.1   *Dynamical Systems*

In the early 1960s, Edward Lorenz[15] studied chaotic phenomena in weather and was among the first to apply high-performance digital computers to solve dynamical systems problems.

More recent work on dynamical systems can trace its heritage to John von Neumann in the 1940s and 1950s, and the works of many on radar and electronics systems of WWII, such as that of Vannevar Bush,[16] who

---

[13] Sir Isaac Newton (1642–1727) was an English mathematician, physicist, astronomer and theologian. Newton made vast contributions to these fields, but is best known for his discovery of what we now call "Newtonian Mechanics," his laws of gravity and motion, and his co-discovery (with Gottfried Leibniz) of Calculus.

[14] Jules Henri Poincaré (1854–1912) was a French mathematician, physicist and engineer. He made many contributions to pure and applied mathematics, mathematical physics and celestial mechanics.

[15] Edward Lorenz (1917–2008) was an American mathematician and meteorologist who pioneered the study of weather and climate, particularly through the introduction of computer modeling. He is considered the founder of modern chaos theory.

[16] Vannevar Bush (1890–1974) was an American scientist, academic, inventor and policy maker. Among his many accomplishments, he was a pioneer in the discipline of systems

also pioneered the use of analog computers in solving system engineering problems.

5.5.2 *Systems Engineering*

The discipline of systems engineering is concerned with the design, implementation, analysis, optimization and management of small- and large-scale interrelated collections of components, people and processes. Systems engineering began to separate from other engineering disciplines such as electrical, industrial and mechanical in the late 1970s. Systems engineering, like other engineering disciplines, is highly mathematical yet practical. Today, you can study systems engineering as a separate discipline at the undergraduate and graduate levels.

5.5.3 *Origins of Systems Thinking*

The modern study of systems thinking began in the 1950s with the work of Jay Forrester at MIT and continued with Russell Ackoff at the University of Pennsylvania in the 1960s and 1970s, and more recently, Peter Senge, also at MIT, and many others. Those wishing to pursue additional concepts of systems thinking would be best off starting with *The Fifth Discipline* (Senge, 1990).

5.6 Exercises and Things to Do

Exercise 5.1

For the following types of systems, give an example. You can describe the system informally or give a function-based definition of the system (e.g. $y = f(t) = t \cdot (t^2 - s)$, which is a nonlinear time-variant system).

(1) linear system
(2) nonlinear system

---

engineering and, in particular, the application of analog computers. Many systems engineering researchers can trace their academic lineage directly to Bush (including Laplante).

(3) time-invariant systems
(4) time-variant system
(5) linear time-variant system
(6) nonlinear time-invariant system
(7) nonlinear time-variant system

Exercise 5.2

Discuss whether a kaleidoscope behaves linearly like the faucet system.

Exercise 5.3

For the system represented by the function $f(r) = r^2$, where r is any real number, show that $f(4 \cdot r + 3 \cdot r) \neq 4 \cdot f(r) + 3 \cdot f(r)$.

Exercise 5.4

Show that the shower faucet system represented by real-valued function $t = f(4)$ is linear.

Exercise 5.5

Show that a system governed by Hooke's Law, $F(x) = -kx$, is linear.

Exercise 5.6

Is a system that can be modeled by the real-valued function $f(x) = 5 \cdot e^x$ linear?

Exercise 5.7

Determine whether the following systems are digital, analog or hybrid. Justify your answer.

(1) a washing machine
(2) an ant colony
(3) an amusement park ride

Exercise 5.8

Investigate the concepts of systems connected in series and in parallel, and write a one-page report. You can describe these informally without the use of mathematics.

Exercise 5.9

Describe a situation in which a significant delay in stimulus–response (or cause and effect) caused some kind of unusual (chaotic-like) behavior.

Exercise 5.10

Besides a door, describe another common system or device that, while it appears to be bistable, is inherently not bistable.

Exercise 5.11

For the pet sitting business described in Section 5.2.1, discuss some possible emergent behaviors if instead of charging extra for late pet pickup, a bonus was paid to the owner for on-time pickup. What other incentives (or disincentives could the pet owner try to ensure on-time pickup? What are some of the possible emergent behaviors for these policies?

Exercise 5.12

Investigate the NASA swarm computing initiative and write a one- to two-page paper summarizing your findings.

Exercise 5.13

Kale is a nutritious, delicious and inexpensive vegetable that can be served in many different ways. We like it in salads, soups, as a side dish and even oven roasted to make kale chips. Find a recipe for kale that you have never tried before and make it to expand your culinary horizons.

Exercise 5.14

Write a short (less than one page) summary of the basic approaches to uncertainty offered by each of the following:

- expert systems
- fuzzy sets and fuzzy logic
- probabilistic reasoning
- neural networks
- rough sets

Exercise 5.15

Experiment with the flock simulator found at https://demonstrations.wolfram.com/BoidsSimulatedFlockingBehavior. Write a one-page paper describing how the various controls change the behavior of the flock/swarm.

Exercise 5.16

Investigate and write a one-page paper summarizing one of the following:

- moving average
- Kalman filter
- sensor fusion

Exercise 5.17

Write a one-page biography for one (or more) of the following individuals:

- Isaac Newton
- Henri Poincaré
- Edward Lorenz

Exercise 5.18

For all or some of the mathematical and scientific figures mentioned in this book, create a historical timeline showing name (birth, death), contribution(s) to chaos, fractals and dynamical systems (year of contribution).

Glossary

A

affine transformations – mathematical operations involving sliding, stretching and rotating.

algorithm – a recipe or set of rules that describe some process.

analog system – a system where all inputs and outputs are in an analog (non-digital) form.

aspect ratio – in computer screens, the ratio of the length of the x-coordinate range to the y-coordinate range.

attractor – the point to which an iterated function tends towards if it does not escape and is not indifferent.

attractor sensitivity – the threshold to which a function $f(z)$ is iterated at the point z_0; if the square of its modulus at any point is less than the attractor sensitivity, then the point attracts.

B

basin of attraction – the set of all points which when iterated by a function f attract to the same point.

bifurcation diagram – a diagram of an iterated function against the value of a swept constant; in many cases, this generates a fractal that tends to have two zones of activity.

binary variables and constants – variables and constants that can only take on the values 0 or 1.

bistable device – a system or device that can be in either of only one of two possible states, but never in both simultaneously.

boolean AND operation – a logical operation on binary variables and constants that produces a one only if both operands are one; it is usually denoted as $\cdot$.

boolean complement – a logical operation on a binary variable or constant that produces a one if the operand is a zero and vice versa; it is usually denoted by a bar over the operand.

boolean OR operation – a logical operation on binary variables and constants that produces a one if one or both operands are one; it is usually denoted as $+$.

C

Cantor middle third argument – a recursive mathematical procedure involving the removal of the middle third of a line segment.

Cantor set – the result of applying the Cantor middle third.

cause-and-effect – an informal way to describe a set of inputs and its associated outputs, *stimulus–response*, for some system.

cellular automata – a type of dynamical system involving matrices or cells.

chaos – a state of disorder.

chaotic system – those that when they are in equilibrium, they are in unstable equilibrium.

complex conjugate – if $z = a + bi$ is a complex number, then its complex conjugate, denoted $\bar{z}$, is $\bar{z} = a - bi$.

complex number – a number that has both real and imaginary parts, for example, in the complex number $3 + 4i$, 3 is the real part and 4 is the imaginary part.

complex plane – a map where complex numbers are plotted; it is similar to the Cartesian plane except that the y-axis is labeled as "iy."

complex variables – placeholders for complex numbers; usually denoted by some variant of the letter z.

compression ratio – the ratio of the bytes required to store an uncompressed image to those needed to store the compressed equivalent; also known as *contact bounce*.

contact bounce – see *switch bounce*.

continuous simulation – a model involving differential equations.

D

daughter wavelets – *wavelets* generated by time shifting and scaling the *mother wavelet* function.

deterministic system – a system where for each possible state and each set of inputs, a unique set of outputs and next state of the system can be determined.

differential equations – an equation involving a function and its derivatives.

digital system – a system where all inputs and outputs are in digital form.

discrete simulation – a model using finite difference equations.

dynamical systems – a sub-field of mathematics that is concerned with the study of phenomena that depend on time; also, dynamical systems represent behaviors that occur based on the repeated application of an algorithm.

E

EKG – see *electrocardiogram*.

electrocardiogram – a graph that depicts the electrical activity in the heart; abbreviated EKG or ECG.

emergent property – a behavior that appears when a number of simple entities operate in an environment, forming more complex behaviors as a collective.

escape – when the result of iterating a function at a point tends toward infinity or minus infinity.

Euclidean space – any three-dimensional space, for example, the one in which we live.

event deterministic system – a system where the next state and output of a system are known for each set of inputs that trigger events; a system that is *deterministic* is also *event deterministic*.

expert system – a form of artificial intelligence in which software program imitates a human expert in some restricted domain of knowledge.

explainable – a property of *expert systems* meaning that the decisions and actions made can be easily explained and are understandable to humans.

F

Fibonacci sequence – a sequence of numbers that begins with 0, 1, and then proceeds by adding the preceding two numbers in a sequence to get the next.

filled Mandelbrot set – a Mandelbrot set where a single color is used.

finite difference equation – a recursive equation that describes a function at time t in terms of the function at previous time samples, $t - 1$, $t - 2$, and so on.

fractal – an image with an infinite amount of self-similarity.

fractal dimension – fractional dimension of geometric images; defined as the logarithm of the number of self-similar pieces divided by the logarithm of the magnification needed to obtain them.

function – in mathematics, a mapping or rule.

function composition – the process of applying a function to result of another function or itself.

function iteration – repeated composition of a function.

fuzzy – a valuation function which provides a mechanism for dealing with the classification of some property that is not "crisply" defined as having or not having that property.

fuzzy logic – a form of Boolean logic in which the truth value of variables may be any real number between 0 and 1.

fuzzy set – a kind of set where membership is not crisply defined as being in or out of the set; rather, a valuation function is assigned with 0 representing it is completely not in the set and 1 representing it is completely in the set.

G

geometry shader – a special program that runs on the GPU (graphics processing unit) and not the CPU; also known as a *shader*.

H

Heisenberg uncertainty – a principle from particle physics that says that the position and speed of a particle cannot both be known precisely at the same time; knowing more about one property means knowing less about the other.

hive mind – aggregate simple behavior manifesting as complex group behavior.

hybrid system – a system where inputs and outputs can be some combination of analog and digital.

hyperbolic cosine – a cosine function defined on real numbers; in particular, if x is a real number, then

$$cosh(x) = \frac{e^x + e^{-x}}{2}$$

hyperbolic sine – a sine function defined on real numbers; in particular, if x is a real number, then

$$sinh(x) = \frac{e^x - e^{-x}}{2}$$

I

image compression – the process of reducing the amount of stored information needed to reproduce an image.

imaginary part – in a complex number, the component that consists of a real number (a number found on the number line) times the positive square root of -1, denoted i.

indifferent – a point that under iteration acts as neither an attractor nor a repelling point.

integers – the set of positive and negative whole numbers, including 0, that is $0, 1, 2, 3, \ldots$ and $-1, -2, -3, \ldots$. Sometimes denoted $\mathbb{Z}$.

inverse – for a real function f, if its inverse is f^{-1}, then $f^{-1}(f(x)) = x$.

iteration – repeated composition of a function or procedure.

iterated function systems – a way to generate fractals by the repeated application of special geometric procedures. Abbreviated as IFS.

J

Julia set – a complex function $f(z)$ is the boundary of the set of points that escape.

K

Kalman filter – a special mathematical approach to *sensor fusion* typically used when the error characteristics of the sources are very different, for example, a Kalman filter could be used to resolve the GPS and inertial measurement data for the position of a spacecraft to obtain a more accurate estimate.

L

linear system – a system which exhibits the properties of linearity, that is, additivity and scaling.

logistics equation – an equation first proposed as a model for population growth; it is given by

$$P(t + 1) = aP(t) - aP(t)^2$$

where $aP(t)$ is the previous year's population plus newborns and $aP(t)^2$ is the death rate; plotting the values of $P(t)$ over many years and over many values of a yields a bifurcation diagram.

M

Mandelbrot set – the set of complex constants c_i for which the orbits of the function

$$f(z) = g(z) + c_i$$

evaluated at the initial condition of $z_0 = 0$ do not escape; the Mandelbrot set is usually found for the function $g(z) = z^2$.

matrix – a mathematical construct that consists of rows and columns that hold numbers.

modulus – of a complex number z, is equal to the square root of the sum of the squares of its real and imaginary parts.

mother wavelet – the base function from which *daughter wavelets* are generated via scaling and shifting.

N

neural network – a type of information processing structure intended to imitate the thinking and behavior of human (and other higher order animal) brains.

non-commutative algebra – an algebraic system where the commutative laws do not hold, for example,

$$x \cdot y = y \cdot x$$

does not hold for quaternions.

non-deterministic system – a system which is not *deterministic*.

nonlinear system – a system which does not exhibit the properties of linearity (see Section 5.1.7).

NP-complete problem – one of a class of computational problems that can only be attacked with brute force approaches; solving any particular problem in the class implies that every NP-complete problem can be solved with a polynomial-time algorithm.

NP-hard problem – one of a class of computational problems that can only be attacked with brute force approaches: there is no known polynomial-time algorithm that solves any single problem in the class.

O

one-dimensional cellular automaton – a cellular automaton, where we trace the evolution of the system by observing a row of cells at time t followed by the row at time $t + 1$ and so on.

OODA loop – OODA is an acronym that stands for "observe," "orient," "decide" and "act," that is, an individual has to be stimulated by the presence of the threat (usually visually, though it could be by any of the other senses), then orient itself to the nature of the potential threat, decide how to respond, and then act accordingly; continuously repeating the sequence acts as a loop between the thought and action.

P

pixel – a screen picture element (point on the screen) capable of displaying one or more colors.

Q

quaternion – a hyper-complex number (pair of complex numbers) used in the generation of three-dimensional fractals; quaternions extend the set of *complex* numbers and are used in many applications, including three-dimensional rendering.

R

random orbits – attracting points determined by the iteration of random starting points using an appropriate geometric procedure.

real number – any floating (or decimal) point number, that is, any number found on a "number line," sometimes denoted as $\mathbb{R}$.

real part – in a complex number, the component that consists of a real number, that is, a number that can be found on the number line.

recursive – mathematical, graphical or geometric procedures that are self-referential.

repelling point – a point that escapes.

resolution – on a computer screen, the density of the pixels.

response – the associated outputs for some set of inputs (stimuli) to a system; informally, this might be referred to as the effect for some associated cause.

response time – for a system, it is the time between the presentation of a set of inputs to a system and the realization of the required behavior, including the availability of all associated outputs.

rough sets – a technique that uses the theory of unifying a collection of differently bounded sets to describe the greater whole.

S

scene analysis – the process of extracting specific features from a larger picture or scene.

second-order effect – the observation that in systems, every action has a consequence; sometimes called *side effect* or unintended consequence.

self-similar – in image, when the structure of the whole is often reflected in every part.

sensitive dependence – on initial conditions, a system that is subject to great variance in later states due to only slight variance in the initial conditions.

sensor fusion – any technique used to combine the information from two or more sensors as a way of resolving the differences in the data or missing data; a *Kalman filter* is one such technique, simple averaging is another.

shader – see *geometry shader*.

Sierpinski gasket/carpet – a fractal created by repeatedly dividing a square into nine equal-sized squares and removing the middle one, also known as Sierpinski carpet.

Sierpinski triangle – a fractal generated by repeatedly dividing a triangle into four self-similar ones and removing the inner fourth one.

stable equilibrium – a system which cannot easily be moved to a chaotic state.

steady state – When a system reaches a point in time in which very little changes or in which every change is balanced by another, it is said to be in a steady state.

stimuli – the set of inputs to a system; a single input is called a *stimulus*. For each stimuli, a set of outputs or *response* is produced.

stimulus–response pair – a way to refer to a set of inputs and its associated outputs together, informally, this might be referred to as a *cause-and-effect*.

strange attractor – when the attracting set of an iterated function or procedures is a fractal.

swarm computing – a biology-inspired approach to decentralized systems composed of relatively simple agents that can self-organize to solve complex problems together through direct or indirect interactions.

switch bounce – the physical phenomenon that a bistable system cannot instantaneously change from one state to another without inducing phantom state changes, also known as *contact bounce*.

system – a mapping function of a set of inputs to a set of outputs, see also *digital system, analog system* and *hybrid* system.

systems engineering – an engineering discipline concerned with the design, implementation, analysis, optimization and management of small- and large-scale interrelated collections of components, people and processes.

systems thinking – a discipline concerned with making sense of complexity by looking at large-scale behaviors, second-order effects, emergent behaviors and interrelationships.

system under control – a system whose inputs are the outputs of another system; there may be outputs from the system under control that are feedback inputs to the controlling system.

T

temporal determinism – in a *deterministic system* when the response time for each set of outputs is known.

time-invariant system – a system in which time does not significantly affect the systems behavior, or, in the case of a theoretical system, not a variable at all.

time-variant system – a system in which time can significantly affect the system's behavior.

tipping point – a point of *unstable equilibrium.*

turbulence – a chaotic system condition characterized by disorder on all scales, with backward eddy currents, and circular waves.

two-dimensional cellular automaton – a cellular automaton where a cell's contents at time t is based on its own and the contents of all its immediate neighbors at time $t - 1$.

U

unstable equilibrium – a system that can easily moved into a chaotic state.

V

vector – a one-dimensional matrix, that is, an ordered list of elements, typically, numbers or variables.

W

wavelet – self-similar functions generated from one basic function called the *mother wavelet*.

Z

Zeno's Paradox – the contradiction that in repeatedly halving the distance from one point to another, one can never reach the end point; conversely, it can be shown mathematically that such a limit can be reached; also known as "Zeno's Achilles Paradox."

Bibliography

A. Antelmi, G. Cordasco, M. D'Auria, D. De Vinco, A. Negro and C. Spagnuolo, On evaluating rust as a programming language for the future of massive agent-based simulations. In *Asian Simulation Conference*, pp. 15–28. Springer, Singapore, 2019.

A. Langlands, L. Titley and O. Nelson, Rust for Visual Effects. In *The Digital Production Symposium*, July 27, 2021, pp. 1–6.

B. J. West and A. L. Goldberger, Physiology in fractal dimensions. *American Scientist*, vol. 75, July–August, 1987, pp. 354–365.

B. Mandlebrot, *The Fractal Geometry of Nature*, W. H. Freeman and Co., New York, 1982.

C. Gargour, M. Gabrea, V. Ramachandran and J. M. Lina, A short introduction to wavelets and their applications. *IEEE Circuits and Systems Magazine*, vol. 9, no. 2, 2009, pp. 57–68.

C. W. Reynolds, Flocks, herds and schools: A distributed behavioral model. In *Proceedings of the 14th annual conference on Computer graphics and interactive techniques*, August 1987, pp. 25–34.

D. R. Hofstadter, *Gödel, Escher, Bach: An Eternal Golden Braid*, 20th Anniversary Edition, Basic Books, New York, 2009.

E. Peters, *Chaos and Order in the Capital Markets, Second Edition*, John-Wiley, Hoboken, NJ, 1996.

F. C. Moon, *Chaotic and Fractal Dynamics*, John Wiley & Sons, New York, 1992.

G. Moran, *Chaos Theory and Psychoanalysis: The Fluidic Nature of the Mind*, Routledge, Oxfordshire, UK, 2018.

H. B. Lin, ed., *Chaos*, World Scientific, Singapore, 1984.

J. Gleick, *Chaos: Making a New Science, Twentieth Anniversary Edition*, Penguin Books, New York, 2008.

J. Kessenich, G. Sellers and D. Shreiner, *OpenGL®Programming Guide: The Official Guide to Learning OpenGL®, Version 4.5 with SPIR-V (9th. ed.).*, Addison-Wesley Professional, 2017.

K. P. Hadeler and J. Müller, *Cellular Automata: Analysis and Applications*, Springer, Cham Switzerland, 2017.

M. Barnsley, *Fractals Everywhere, New Edition*, Dover Books, Mineola, NY, 2012.

M. C. Escher, *Escher on Escher: Exploring the Infinite*, H. N. Abrams, Inc., Publishers, New York, NY, 1989.

P. M. Senge, *The Fifth Discipline*, Doubleday/Currency, New York, 1990.

S. Wolfram, ed., *A New Kind of Science*, Wolfram Media, Champaign, IL, 2002.

T. DeAngelis, Chaos, chaos everywhere is what the theorists think. *The American Psychological Association Monitor*, vol. 24, no. 1, January, 1993, pp. 1, 41.

Index

A

Achilles Paradox, 29
addition, complex numbers, 41
affine transformations, 18, 25
algebra, non-commutative, 67
algorithms, 6, 72, 139, 157
  histogram coloring, 52
  origin of the word, 6
alveoli, 100
analog systems, 130
artificial intelligence, 151
attracting points, 7
  attractor sensitivity, 47
  complex numbers, Julia sets, 46
  Mandelbrot sets, 61
attractors, 8
  bifurcation diagrams, 9
  strange attractors, 9

B

baklava, 5
Barnsley, Michael, 23, 35, 81, 112
basin of attraction, boundary scanning
  method (BSM), 66
behavior, chaos of the mind, 103
bifurcation diagrams, economic systems
  simulation, 117
  neuron growth patterns, 100
bifurcation.rs program, 10
  GLSL code, 12
  graphic output, 10, 117

price mode, 117
  source code, 11
Binet, Jacques, 27
Bonacci, Leonardo (Fibonacci), 26
Boolean logic, 120
boundary scanning method (BSM), 46,
  65
Brewster, David, 132
bronchial growth patterns, 100
bronchial tubes, 100
BSM, basin of attraction, 66
  Julia sets, 46
  Mandelbrot sets, 61
Bush, Vannevar, 160

C

Cantor sets, 28
  fractal dimension, 31
  middle third argument, 28
Cantor, Georg, 28
cantor.rs program, 29
  graphic output, 28
  source code, 30
carpet, *see* Sierpinski carpet, 18
carpet-ifs.rs program, 23
  graphic output, 24
  IFS codes, 23
castle.rs program, 108
  graphics output, 109
  IFS codes, 109
cause-and-effect, 132

cellular automata, 34, 118
  Boolean logic, 120
  Game of Life, see Game of Life, 124
  one-dimensional, 119
  self-organizing, 123
  Sierpinski triangle, 121
  swarms, 157
  two-dimensional, 124
  von Neumann machines, 118
  Wolfram classifications, 119
chaos, 2
  fractals and chaos, 33
  history of chaotic theory, 34
  infinite rollercoaster concept, 2
  sensitive dependence, 2
chrysanthemum *see also* mandel_julia.rs
  program, flower2, 88
clouds, 90
  Julia set, 92
clouds.rs program, graphic output, 93
  IFS codes, 93
coastline, self-similarity, 3
coastlines, natural chaos and fractals, 98
collage theorem, 112
complex numbers and functions, 39
  addition, 41
  arithmetic with complex numbers, 41
  attractors of complex numbers, 46
  complex conjugate, 43
  complex plane, 41
  cosine, hyperbolic cosine, 44
  division, 42
  Euler's equation, 45
  exponential, 45
  GLSL procedures, 42–45
  imaginary part, 40
  modulus, 47
  multiplication, 42
  non-commutative algebra, 67
  plotting, 40
  quaternions, 67
  real part, 40
  sine, hyperbolic sine, 44
  subtraction, 41
  three-dimensional fractals, 67
  variables, complex variables, 40

composition, function composition, 7
compound interest, 134
compression, 111
compression ratio, 112
continuous simulation, 77
control systems, 136
Conway, John, 124
cosines, hyperbolic, complex numbers,
  44
cross-fractals, 96
  IFS codes, 98
crystal growth, 122

D

1d_life.rs program, 120
  graphic output, 121
  input syntax, 120, 123
  rule 2, 122
  rule 2, graphic output, 123
DeAngelis, 100
declaration, function declaration, 49
definition, function definition, 49
dendrites, 101
Devaney, Robert, 35
differential equation, 77
digital systems, 130
dimension, fractal dimension, 31
discrete simulation, 77
division, complex numbers, 42
Douady, Adrien, 50
Douady's rabbit, 53–57
Douady's rabbit (Julia set), GLSL
  subroutine, 51
Douady's rabbit, *see* julia_simple.rs
  program, Douady's rabbit, 50
dragon, *see* mandel_julia.rs program,
  dragon, 58
dynamical systems, 34

E

economic systems, 115
  logistics, 117
EKG, Julia set, *see* mandel_julia.rs
  program, EKG, 102
  wavelets, 70

emergent behavior, simple example, 146
emergent property, 145
equilibrium, stable vs. unstable systems, 1
escaping orbits, 46
escaping points, 7
Escher, M. C., 28
Euclidean space, 29
expert systems, 151
   explainable, 152
exponent, 32
exponential notation, 7
   complex numbers, 45
exponential numbers, Euler's equation, 45

F

fall.rs program, 96
   graphic output, 97
   IFS codes, 98
fern-ifs.rs program, 82
   graphic output, 82
   IFS codes, 83
ferns, 81
fibo.rs program, 27
Fibonacci sequence, 26
   Binet formula, 27
   Rust code, 27
fiddle head fern, 81, 105
floating-point number, 17
flowers (Julia sets), 87
   graphic output, 90–91
forest.rs program, 83
   graphic output, 85
   source code, 86
fractals everywhere, 23
fractals, bifurcation diagrams, 10
   Cantor sets, 28
   human body, 99
   human-made, 108
   in nature, 78
   iterated function systems (IFS), 13
   Julia sets, 46
   Mandelbrot set, 61
   recursive generation of fractals, 25
   Siegel disk, 57

Sierpinski triangle, 13
   three-dimensional fractals, 67
functions, 7
   composition, 7
   iteration, 7
fuzzy logic, 153
fuzzy sets, 153
fuzzy values, 153

G

galaxies, 96
galaxy1.rs program, 97
   graphic output, 99
Game of Life, blinkers, 125
   other implementations, 126
game_of_life.rs program, 124
   graphic output, 125
   graphic output, stable and bi-stable formations, 126
gamut, 55
gasket, *see* Sierpinski carpet, 18
genetics, 79
GLSL, subroutines, 49
growth systems, 133

H

Hamilton, William Rowan, 67
Heaviside, Oliver, 68
Heisenberg uncertainty, 80, 149
Heisenberg, Werner, 149
histogram, 55
hive mind, 156
honeybees, 156
Hooke's Law, 142
Hooke, Robert, 142
human body, fractals, 99

I

IFS, reading IFS code tables, 20
imaginary part of complex numbers, 40
indifferent points, 8
inferno color scheme, 52
interpolation, 56
inverse iteration method (IIM), 46, 65

iterated function systems (IFS), 18
iteration, function iteration, 7

J

Julia, Gaston, 34
julia_simple.rs program, 48
  cos, 49
  cos, GLSL subroutine, 49
  cos, graphic output, 50
  Douady's rabbit, graphic output, 51
  fragment shader GLSL code, 48

K

al-Khwarizmi, Mohammad ibn Musa, 6
Kalman filtering, 155
kinematics, 67

L

limits, 29, 135
linear growth, 133
linear interpolation, 56
linear systems, 140
linked lists, 11
logarithm, 32
logistics, 117
Lorenz, Edward, 160

M

mandel_julia.rs program, amoeba, graphics output, 80
  amoeba, 79
  cloud, colorizing, 92
  cloud, graphic output, 91
  dendrite, 101
  dendrite, graphic output, 101
  dragon, GLSL subroutine, 59
  dragon, graphic output, 59
  EKG, 102
  EKG, graphic output, 102
  flower1, 87
  flower1, graphic output, 90
  flower2, 88
  flower2, graphic output, 91
  Mandelbrot mode, 65

Mandelbrot mode, graphic output, 65
  sin, GLSL subroutine, 60
  sin, graphic output, 60
  snowflakes, 95
  snowflakes, colorization, 96
  snowflakes, graphic output, 95
  rabbit, 55
  Siegel disk, GLSL subroutine, 58
  Siegel disk, graphic output, 58
Mandelbrot, Benoit, 61
Mandelbrot set, 3
  filled, 62
mandelbrot_simple.rs program, 62
  filled Mandelbrot set, 63
  fragment shader GLSL code, 62
  graphic output, 63
matrix, matrices, 19
Maxwell's equations, 68
Maxwell, James Clerk, 68
maze.rs program, 110
  graphic output, 110
  IFS codes, 111
Mexican hat function, 69
minimalism, 5
modeling, as systems, 131
multiplication, complex numbers, 42

N

neural networks, 154
Newton, Isaac, 160
nonlinear systems, 143
NP-complete, 148
NP-hard, 148

O

observe, orient, decide, act (OODA) loops, 138
octiles, 56

P

Pascal, Blaise, 26
Peitgen, Heinz-Otto, 35
physiological processes, 102
Poincaré, Jules Henri, 160

probabilistic reasoning, 152
pseudo-code, 6

Q

quantiles, 56

R

real part of complex numbers, 40
recursion, 25, 27
redmoscl.rs program, 87
  graphic output, 88
redwood forest, (*see* redmoscl.rs
  program), 87
repelling points, *see* escaping points, 8
rocks and boulders, 92
Romanesco, 81, 105
rough sets, 154

S

scene analysis, 111
seals.rs program, 78
  graphic output, 78
  IFS codes, 79
seaweed.rs program, 87
  graphic output, 89
  IFS codes, 90
self-similarity, 3, 70, 116
sensor fusion, 155
sfogliatelle, 5
Siegel, Carl Ludwig, 57
Sierpinski triangle, *see also* sierpinski.rs
  program, sierpinski-ifs.rs program),
  13
Sierpinski triangle, 1d_life.rs output, 121
  IFS procedure visualization, 19
  mapping procedure, 16
  random orbits, 15
sierpinski-ifs.rs program, 22
  IFS codes, 20
  source code, 22
sierpinski.rs program, 14
  graphic output, 14
  source code, 17
Sieve of Eratosthenes, 7, 36
sines, hyperbolic, complex numbers, 44

Small Wonder Honey, 156
snowflakes, 95
spring constant, 142
square root, 39
state uncertainty, 148
subtraction, complex numbers, 41
supply chains, 138
swamp.rs program, 114
  graphic output, 114
  IFS codes, 115
swarm computing, 159
swarm simulator, 157
swarms, 155
  applications, 158
Swift, Jonathan, 25
system theory, 130
systems, 130
  analog systems, 130
  bistable systems, 149
  determinism, 139
  digital systems, 130
  dynamical systems, 5
  emergent behaviors, 145
  event determinism, 139
  feedback, 137
  growth, 133
  history of dynamical systems and
    systems theory, 160
  hybrid systems, 130
  linear systems, 140
  linear systems in nature, 142
  modeling, 131
  nonlinear systems, 143
  responses, 132
  response time, 136
  second-order effects, 147
  steady state, 132
  stimuli, 132
  summary of system types, 144
  systems engineering, 161
  systems thinking, 144
  temporal determinism, 139
  time-variant vs. time-invariant,
    140
  unstable vs. stable systems, 1

T

time-variant vs. time-invariant, 140
tipping point, 1
tree-ifs.rs program, 83
    graphic output, 84
    IFS codes, 83
turbulence, 107
turbulent flow, 107

U

uncertainty, 148
    and chaos, 150
    dealing with uncertainty, 150
    expert systems, 151
    fuzzy sets and logic, 153
    multiple sources of uncertainty, 155
    neural networks, 154

probabilistic reasoning, 152
    rough sets, 154
    state uncertainty, 148
unstable equilibrium, 1

V

Vec, 11
vectors, 11
von Neumann, John, 118

W

Wacław, Sierpiński, 18
wavelets, 68
    daughter wavelets, 68
    mother wavelet, 68
weather, 79
Wolfram, Stephen, 119